Design Concepts
for Engineers

Mark N. Horenstein

Boston University

Prentice Hall
Upper Saddle River, NJ 07458

Library of Congress Information Available

Editor-in-chief: **MARCIA HORTON**
Acquisitions editor: **ERIC SVENDSEN**
Director of production and manufacturing: **DAVID W. RICCARDI**
Managing editor: **EILEEN CLARK**
Editorial/production supervision: **ROSE KERNAN**
Cover director: **JAYNE CONTE**
Creative director: **AMY ROSEN**
Marketing manager: **DANNY HOYT**
Manufacturing buyer: **PAT BROWN**
Editorial assistant: **GRIFFIN CABLE**

Printed in the United States of America

10 9 8 7 6 5 4 3 2 1

ISBN 0-13-081369-9

Prentice-Hall International (UK) Limited, *London*
Prentice-Hall of Australia Pty. Limited, *Sydney*
Prentice-Hall Canada, Inc., *Toronto*
Prentice-Hall Hispanoamericana, S.A., *Mexico*
Prentice-Hall of India Private Limited, *New Delhi*
Prentice-Hall of Japan, Inc., *Tokyo*
Prentice-Hall (Singapore) Pte., Ltd., *Singapore*
Editora Prentice-Hall do Brazil, Ltda., *Rio de Janeiro*

About ESource

The Challenge

Professors who teach the Introductory/First-Year Engineering course popular at most engineering schools have a unique challenge—teaching a course defined by a changing curriculum. The first-year engineering course is different from any other engineering course in that there is no real cannon that defines the course content. It is not like Engineering Mechanics or Circuit Theory where a consistent set of topics define the course. Instead, the introductory engineering course is most often defined by the creativity of professors and students, and the specific needs of a college or university each semester. Faculty involved in this course typically put extra effort into it, and it shows in the uniqueness of each course at each school.

Choosing a textbook can be a challenge for unique courses. Most freshmen require some sort of reference material to help them through their first semesters as a college student. But because faculty put such a strong mark on their course, they often have a difficult time finding the right mix of materials for their course and often have to go without a text, or with one that does not really fit. Conventional textbooks are far too static for the typical specialization of the first-year course. How do you find the perfect text for your course that will support your students educational needs, but give you the flexibility to maximize the potential of your course?

ESource—The Prentice Hall Engineering Source
http://emissary.prenhall.com/esource

Prentice Hall created ESource—The Prentice-Hall Engineering Source—to give professors the power to harness the full potential of their text and their freshman/first year engineering course. In today's technologically advanced world, why settle for a book that isn't perfect for your course? Why not have a book that has the exact blend of topics that you want to cover with your students?

More then just a collection of books, ESource is a unique publishing system revolving around the ESource website—http://emissary.prenhall.com/esource/. ESource enables you to put your stamp on your book just as you do your course. It lets you:

Control You choose exactly what chapters or sections are in your book and in what order they appear. Of course, you can choose the entire book if you'd like and stay with the author's original order.

Optimize Get the most from your book and your course. ESource lets you produce the optimal text for your student's needs.

Customize   You can add your own material anywhere in your text's presentation, and your final product will arrive at your bookstore as a professionally formatted text.

ESource Content

All the content in ESource was written by educators specifically for freshman/first-year students. Authors tried to strike a balanced level of presentation, one that was not either too formulaic and trivial, but not focusing heavily on advanced topics that most introductory students will not encounter until later classes. A developmental editor reviewed the books and made sure that every text was written at the appropriate level, and that the books featured a balanced presentation. Because many professors do not have extensive time to cover these topics in the classroom, authors prepared each text with the idea that many students would use it for self-instruction and independent study. Students should be able to use this content to learn the software tool or subject on their own.

While authors had the freedom to write texts in a style appropriate to their particular subject, all followed certain guidelines created to promote the consistency a text needs. Namely, every chapter opens with a clear set of objectives to lead students into the chapter. Each chapter also contains practice problems that tests a student's skill at performing the tasks they have just learned. Chapters close with extra practice questions and a list of key terms for reference. Authors tried to focus on motivating applications that demonstrate how engineers work in the real world, and included these applications throughout the text in various chapter openers, examples, and problem material. Specific Engineering and Science **Application Boxes** are also located throughout the texts, and focus on a specific application and demonstrating its solution.

Because students often have an adjustment from high school to college, each book contains several **Professional Success Boxes** specifically designed to provide advice on college study skills. Each author has worked to provide students with tips and techniques that help a student better understand the material, and avoid common pitfalls or problems first-year students often have. In addition, this series contains an entire book titled *Engineering Success* by Peter Schiavone of the University of Alberta intended to expose students quickly to what it takes to be an engineering student.

Creating Your Book

Using ESource is simple. You preview the content either on-line or through examination copies of the books you can request on-line, from your PH sales rep, or by calling(1-800-526-0485). Create an on-line outline of the content you want in the order you want using ESource's simple interface. Either type or cut and paste your own material and insert it into the text flow. You can preview the overall organization of the text you've created at anytime (please note, since this preview is immediate, it comes unformatted.), then press another button and receive an order number for your own custom book . If you are not ready to order, do nothing—ESource will save your work. You can come back at any time and change, re-arrange, or add more material to your creation. You are in control. Once you're finished and you have an ISBN, give it to your bookstore and your book will arrive on their shelves six weeks after the order. Your custom desk copies with their instructor supplements will arrive at your address at the same time.

To learn more about this new system for creating the perfect textbook, go to **http://emissary.prenhall.com/esource/**. You can either go through the on-line walkthrough of how to create a book, or experiment yourself.

Community

ESource has two other areas designed to promote the exchange of information among the introductory engineering community, the Faculty and the Student Centers. Created and maintained with the help of Dale Calkins, an Associate Professor at the University of Washington, these areas contain a wealth of useful information and tools. You can preview outlines created by other schools and can see how others organize their courses. Read a monthly article discussing important topics in the curriculum. You can post your own material and share it with others, as well as use what others have posted in your own documents. Communicate with our authors about their books and make suggestions for improvement. Comment about your course and ask for information from others professors. Create an on-line syllabus using our custom syllabus builder. Browse Prentice Hall's catalog and order titles from your sales rep. Tell us new features that we need to add to the site to make it more useful.

Supplements

Adopters of ESource receive an instructor's CD that includes solutions as well as professor and student code for all the books in the series. This CD also contains approximately **350 Powerpoint Transparencies** created by Jack Leifer—of University South Carolina—Aiken. Professors can either follow these transparencies as pre-prepared lectures or use them as the basis for their own custom presentations. In addition, look to the web site to find materials from other schools that you can download and use in your own course.

Titles in the ESource Series

Introduction to Unix
0-13-095135-8
David I. Schwartz

Introduction to Maple
0-13-095133-1
David I. Schwartz

Introduction to Word
0-13-254764-3
David C. Kuncicky

Introduction to Excel
0-13-254749-X
David C. Kuncicky

Introduction to MathCAD
0-13-937493-0
Ronald W. Larsen

Introduction to AutoCAD, R. 14
0-13-011001-9
Mark Dix and Paul Riley

Introduction to the Internet, 2/e
0-13-011037-X
Scott D. James

Design Concepts for Engineers
0-13-081369-9
Mark N. Horenstein

Engineering Design—A Day in the Life of Four Engineers
0-13-660242–8
Mark N. Horenstein

Engineering Ethics
0-13-784224-4
Charles B. Fleddermann

Engineering Success
0-13-080859-8
Peter Schiavone

Mathematics Review
0-13-011501-0
Peter Schiavone

Introduction to C
0-13-011854-0
Delores M. Etter

Introduction to C++
0-13-011855-9
Delores M. Etter

Introduction to MATLAB
0-13-013149-0
Delores M. Etter and David C. Kuncicky

Introduction to FORTRAN 90
0-13-013146-6
Larry Nyhoff and Sanford Leestma

About the Authors

No project could ever come to pass without a group of authors who have the vision and the courage to turn a stack of blank paper into a book. The authors in this series worked diligently to produce their books, provide the building blocks of the series.

Delores M. Etter is a Professor of Electrical and Computer Engineering at the University of Colorado. Dr. Etter was a faculty member at the University of New Mexico and also a Visiting Professor at Stanford University. Dr. Etter was responsible for the Freshman Engineering Program at the University of New Mexico and is active in the Integrated Teaching Laboratory at the University of Colorado. She was elected a Fellow of the Institute of Electrical and Electronic Engineers for her contributions to education and for her technical leadership in digital signal processing. IN addition to writing best-selling textbooks for engineering computing, Dr. Etter has also published research in the area of adaptive signal processing.

Sanford Leestma is a Professor of Mathematics and Computer Science at Calvin College, and received his Ph.D from New Mexico State University. He has been the long time co-author of successful textbooks on Fortran, Pascal, and data structures in Pascal. His current research interests are in the areas of algorithms and numerical computation.

Larry Nyhoff is a Professor of Mathematics and Computer Science at Calvin College. After doing bachelors work at Calvin, and Masters work at Michigan, he received a Ph.D. from Michigan State and also did graduate work in computer science at Western Michigan. Dr. Nyhoff has taught at Calvin for the past 34 years—mathematics at first and computer science for the past several years. He has co-authored several computer science textbooks since 1981 including titles on Fortran and C++, as well as a brand new title on Data Structures in C++.

Acknowledgments: We express our sincere appreciation to all who helped in the preparation of this module, especially our acquisitions editor Alan Apt, managing editor Laura Steele, development editor Sandra Chavez, and production editor Judy Winthrop. We also thank Larry Genalo for several examples and exercises and Erin Fulp for the Internet address application in Chapter 10. We appreciate the insightful review provided by Bart Childs. We thank our families—Shar, Jeff, Dawn, Rebecca, Megan, Sara, Greg, Julie, Joshua, Derek, Tom, Joan; Marge, Michelle, Sandy, Lori, Michael—for being patient and understanding. We thank God for allowing us to write this text.

Mark Dix began working with AutoCAD in 1985 as a programmer for CAD Support Associates, Inc. He helped design a system for creating estimates and bills of material directly from AutoCAD drawing databases for use in the automated conveyor industry. This system became the basis for systems still widely in use today. In 1986 he began collaborating with Paul Riley to create AutoCAD training materials, combining Riley's background in industrial design and training with Dix's background in writing, curriculum development, and programming. Dix and Riley have created tutorial and teaching methods for every AutoCAD release since Version 2.5. Mr. Dix has a Master of Arts in Teaching from Cornell University and a Masters of Education from the University of Massachusetts. He is currently the Director of Dearborn Academy High School in Arlington, Massachusetts.

Paul Riley is an author, instructor, and designer specializing in graphics and design for multimedia. He is a founding partner of CAD Support Associates, a contract service and professional training organization for computer-aided design. His 15 years of business experience and 20 years of teaching experience are supported by degrees in education and computer science. Paul has taught AutoCAD at the University of Massachusetts at Lowell and is presently teaching AutoCAD at Mt. Ida College in Newton, Massachusetts. He has developed a program, Computer-Aided Design for Professionals that is highly regarded by corporate clients and has been an ongoing success since 1982.

David I. Schwartz is a Lecturer at SUNY-Buffalo who teaches freshman and first-year engineering, and has a Ph.D from SUNY-Buffalo in Civil Engineering. Schwartz originally became interested in Civil engineering out of an interest in building grand structures, but has also pursued other academic interests including artificial intelligence and applied mathematics. He became interested in Unix and Maple through their application to his research, and eventually jumped at the chance to teach these subjects to students. He tries to teach his students to become incremental learners and encourages frequent practice to master a subject, and gain the maturity and confidence to tackle other subjects independently. In his spare time, Schwartz is an avid musician and plays drums in a variety of bands.

Acknowledgments: I would like to thank the entire School of Engineering and Applied Science at the State University of New York at Buffalo for the opportunity to teach not only my students, but myself as well; all my EAS140 students, without whom this book would not be possible—thanks for slugging through my lab packets; Andrea Au, Eric Svendsen, and Elizabeth Wood at Prentice Hall for advising and encouraging me as well as wading through my blizzard of e-mail; Linda and Tony for starting the whole thing in the first place; Rogil Camama, Linda Chattin, Stuart Chen, Jeffrey Chottiner, Roger Christian, Anthony Dalessio, Eugene DeMaitre, Dawn Halvorsen, Thomas Hill, Michael Lamanna, Nate "X" Patwardhan, Durvejai Sheobaran, "Able" Alan Somlo, Ben Stein, Craig Sutton, Barbara Umiker, and Chester "JC" Zeshonski for making this book a reality; Ewa Arrasjid, "Corky" Brunskill, Bob Meyer, and Dave Yearke at "the Department Formerly Known as ECS" for all their friendship, advice, and respect; Jeff, Tony, Forrest, and Mike for the interviews; and, Michael Ryan and Warren Thomas for believing in me.

Ronald W. Larsen is an Associate Professor in Chemical Engineering at Montana State University, and received his Ph.D from the Pennsylvania State University. Larsen was initially attracted to engineering because he felt it was a serving profession, and because engineers are often called on to eliminate dull and routine tasks. He also enjoys the fact that engineering rewards creativity and presents constant challenges. Larsen feels that teaching large sections of students is one of the most challenging tasks he has ever encountered because it enhances the importance of effective communication. He has drawn on a two year experince teaching courses in Mongolia through an interpreter to improve his skills in the classroom. Larsen sees software as one of the changes that has the potential to radically alter the way engineers work, and his book Introduction to Mathcad was written to help young engineers prepare to be productive in an ever-changing workplace.

Acknowledgments: To my students at Montana State University who have endured the rough drafts and typos, and who still allow me to experiment with their classes—my sincere thanks.

Peter Schiavone is a professor and student advisor in the Department of Mechanical Engineering at the University of Alberta. He received his Ph.D. from the University of Strathclyde, U.K. in 1988. He has authored several books in the area of study skills and academic success as well as numerous papers in scientific research journals.

Before starting his career in academia, Dr. Schiavone worked in the private sector for Smith's Industries (Aerospace and Defence Systems Company) and Marconi Instruments in several different areas of engineering including aerospace, systems and software engineering. During that time he developed an interest in engineering research and the applications of mathematics and the physical sciences to solving real-world engineering problems.

His love for teaching brought him to the academic world. He founded the first Mathematics Resource Center at the University of Alberta: a unit designed specifically to teach high school students the necessary survival skills in mathematics and the physical sciences required for first-year engineering. This led to the Students' Union Gold Key award for outstanding contributions to the University and to the community at large.

Dr. Schiavone lectures regularly to freshman engineering students, high school teachers, and new professors on all aspects of engineering success, in particular, maximizing students' academic performance. He wrote the book *Engineering Success* in order to share with you the *secrets of success in engineering study*: the most effective, tried and tested methods used by the most successful engineering students.

Acknowledgments: I'd like to acknowledge the contributions of: Eric Svendsen, for his encouragement and support; Richard Felder for being such an inspiration; the many students who shared their experiences of first-year engineering—both good and bad; and finally, my wife Linda for her continued support and for giving me Conan.

Scott D. James is a staff lecturer at Kettering University (formerly GMI Engineering & Management Institute) in Flint, Michigan. He is currently pursuing a Ph.D. in Systems Engineering with an emphasis on software engineering and computer-integrated manufacturing. Scott decided on writing textbooks after he found a void in the books that were available. "I really wanted a book that showed how to do things in good detail but in a clear and concise way. Many of the books on the market are full of fluff and force you to dig out the really important facts." Scott decided on teaching as a profession after several years in the computer industry. "I thought that it was really important to know what it was like outside of academia. I wanted to provide students with classes that were up to date and provide the information that is really used and needed."

Acknowledgments: Scott would like to acknowledge his family for the time to work on the text and his students and peers at Kettering who offered helpful critique of the materials that eventually became the book.

David C. Kuncicky is a native Floridian. He earned his Baccalaureate in psychology, Master's in computer science, and Ph.D. in computer science from Florida State University. Dr. Kuncicky is the Director of Computing and Multimedia Services for the FAMU-FSU College of Engineering. He also serves as a faculty member in the Department of Electrical Engineering. He has taught computer science and computer engineering courses for the past 15 years. He has published research in the areas of intelligent hybrid systems and neural networks. He is actively involved in the education of computer and network system administrators and is a leader in the area of technology-based curriculum delivery.

Acknowledgments: Thanks to Steffie and Helen for putting up with my late nights and long weekends at the computer. Thanks also to the helpful and insightful technical reviews by the following people: Jerry Ralya, Kathy Kitto of Western Washington University, Avi Singhal of Arizona State University, and Thomas Hill of the State University of New York at Buffalo. I appreciate the patience of Eric Svendsen and Rose Kernan of Prentice Hall for gently guiding me through this project. Finally, thanks to Dean C.J. Chen for providing continued tutelage and support.

Mark Horenstein is an Associate Professor in the Electrical and Computer Engineering Department at Boston University. He received his Bachelors in Electrical Engineering in 1973 from Massachusetts Institute of Technology, his Masters in Electrical Engineering in 1975

from University of California at Berkeley, and his Ph.D. in Electrical Engineering in 1978 from Massachusetts Institute of Technology. Professor Horenstein's research interests are in applied electrostatics and electromagnetics as well as microelectronics, including sensors, instrumentation, and measurement. His research deals with the simulation, test, and measurement of electromagnetic fields. Some topics include electrostatics in manufacturing processes, electrostatic instrumentation, EOS/ESD control, and electromagnetic wave propagation.

Professor Horenstein designed and developed a class at Boston University, which he now teaches entitled Senior Design Project (ENG SC 466). In this course, the student gets real engineering design experience by working for a virtual company, created by Professor Horenstein, that does real projects for outside companies—almost like an apprenticeship. Once in "the company" (Xebec Technologies), the student is assigned to an engineering team of 3-4 persons. A series of potential customers are recruited, from which the team must accept an engineering project. The team must develop a working prototype deliverable engineering system that serves the need of the customer. More than one team may be assigned to the same project, in which case there is competition for the customer's business.

Acknowledgements: Several individuals contributed to the ideas and concepts presented in Design Principles for Engineers. The concept of the Peak Performance design competition, which forms a cornerstone of the book, originated with Professor James Bethune of Boston University. Professor Bethune has been instrumental in conceiving of and running Peak Performance each year and has been the inspiration behind many of the design concepts associated with it. He also provided helpful information on dimensions and tolerance. Several of the ideas presented in the book, particularly the topics on brainstorming and teamwork, were gleaned from a workshop on engineering design help bi-annually by Professor Charles Lovas of Southern Methodist University. The principles of estimation were derived in part from a freshman engineering problem posed by Professor Thomas Kincaid of Boston University.

I would like to thank my family, Roxanne, Rachel, and Arielle, for giving me the time and space to think about and write this book. I also appreciate Roxanne's inspiration and help in identifying examples of human/machine interfaces.

Dedicated to Roxanne, Rachel, and Arielle

 **Charles B. Fleddermann** is a professor in the Department of Electrical and Computer Engineering at the University of New Mexico in Albuquerque, New Mexico. He is a third generation engineer—his grandfather was a civil engineer and father an aeronautical engineer—so "engineering was in my genetic makeup." The genesis of a book on engineering ethics was in the ABET requirement to incorporate ethics topics into the undergraduate engineering curriculum. "Our department decided to have a one-hour seminar course on engineering ethics, but there was no book suitable for such a course." Other texts were tried the first few times the course was offered, but none of them presented ethical theory, analysis, and problem solving in a readily accessible way. "I wanted to have a text which would be concise, yet would give the student the tools required to solve the ethical problems that they might encounter in their professional lives."

Reviewers

ESource benefited from a wealth of reviewers who on the series from its initial idea stage to its completion. Reviewers read manuscripts and contributed insightful comments that helped the authors write great books. We would like to thank everyone who helped us with this project.

Concept Document
Naeem Abdurrahman- University of Texas, Austin
Grant Baker- University of Alaska, Anchorage
Betty Barr- University of Houston
William Beckwith- Clemson University
Ramzi Bualuan- University of Notre Dame
Dale Calkins- University of Washington
Arthur Clausing- University of Illinois at Urbana-Champaign
John Glover- University of Houston
A.S. Hodel- Auburn University
Denise Jackson- University of Tennessee, Knoxville
Kathleen Kitto- Western Washington University
Terry Kohutek- Texas A&M University
Larry Richards- University of Virginia
Avi Singhal- Arizona State University
Joseph Wujek- University of California, Berkeley
Mandochehr Zoghi- University of Dayton

Books
Stephen Allan- Utah State University
Naeem Abdurrahman - University of Texas Austin
Anil Bajaj- Purdue University
Grant Baker - University of Alaska - Anchorage
Betty Barr - University of Houston

William Beckwith - Clemson University
Haym Benaroya- Rutgers University
Tom Bledsaw- ITT Technical Institute
Tom Bryson- University of Missouri, Rolla
Ramzi Bualuan - University of Notre Dame
Dan Budny- Purdue University
Dale Calkins - University of Washington
Arthur Clausing - University of Illinois
James Devine- University of South Florida
Patrick Fitzhorn - Colorado State University
Dale Elifrits- University of Missouri, Rolla
Frank Gerlitz - Washtenaw College
John Glover - University of Houston
John Graham - University of North Carolina-Charlotte
Malcom Heimer - Florida International University
A.S. Hodel - Auburn University
Vern Johnson- University of Arizona
Kathleen Kitto - Western Washington University
Robert Montgomery- Purdue University
Mark Nagurka- Marquette University
Ramarathnam Narasimhan- University of Miami
Larry Richards - University of Virginia
Marc H. Richman - Brown University
Avi Singhal-Arizona State University
Tim Sykes- Houston Community College
Thomas Hill- SUNY at Buffalo
Michael S. Wells - Tennessee Tech University
Joseph Wujek - University of California - Berkeley
Edward Young- University of South Carolina
Mandochehr Zoghi - University of Dayton

Contents

7 EFFECTIVE COMMUNICATION 154

1

What Is Engineering?

If you're reading this book, you're probably enrolled in an introductory course in engineering. You may have chosen engineering because of your strong skills in science and mathematics. Perhaps you like to tinker with things or use computers. Whatever your reason for studying engineering, you are about to embark on a journey that will be full of excitement, discovery, and creativity. Imagine yourself several years from now, after you've finished your college studies in engineering. What will life be like as an engineer? How will everything you learned in school relate to your work and your career? If you plan to become an engineer, these questions are important ones for you to answer. This book will provide you with a vision of the future while helping to teach you the important principles of engineering design.

As an aspiring engineer, you have much to learn. You must master basic mathematics, physics, and chemistry, because these subjects form the foundation for all engineering disciplines. You must study specialized subjects, such as circuits and mechanics, because these courses will allow you to specialize in your chosen discipline. You also must develop an ability to stay on top of technological advances through a program of lifelong learning, because the world embraces new technology almost on a daily basis. Many of

OBJECTIVES

In this chapter, you will learn about:

- Engineering as a career.
- The relationship between the engineer and other professionals.
- Engineering professional organizations.
- Knowledge, experience, and intuition as the foundations of engineering design.

your college courses will provide you with the knowledge and analytical skills that you'll need to function in the engineering world, but you also must learn about the primary focus of the engineer: the practice of design. The ability to build real things is what sets the engineer apart from professionals in the basic sciences. While physicists, chemists, and biologists examine the world and draw general conclusions by observing specific phenomena, the engineer moves *from* the general *to* the specific. The engineer harnesses the laws of nature and utilizes them to produce devices or systems that perform tasks and solve problems. This process defines the essence of design, and you must become proficient at it if you want to become an engineer. This book will teach you the principles of design and help you to apply them to your class assignments, design projects, and future job activities.

1.1 THE MANY FIELDS OF ENGINEERING

A perusal of catalogues from engineering colleges around the world will reveal a large variety of engineering programs. Although the names may vary slightly, most engineers come from one of the following traditional engineering degree programs (listed alphabetically with no preference implied): aeronautical, agricultural, biomedical, chemical, civil, computer, electrical, industrial, mechanical, naval, petroleum, and systems. From reading this list, one might get the impression that engineers are highly specialized, segregated professionals who have little interaction with people from other fields. In truth, engineers tend to be multidisciplinary individuals who are familiar with many different fields. The mechanical engineer knows something about electrical circuits, and the electrical engineer understands basic mechanics. The computer engineer is familiar with the algorithms of industrial processes, and the industrial engineer knows how to program computers. Many of the great engineering accomplishments of the past century, including our global communication and transportation networks, the microelectronic revolution, life-extending biomedical technology, and inexpensive, reliable air transportation, were made possible by interactive teams of engineers from a multitude of disciplines.

Although engineers are multidisciplanary in nature, most are trained in a specific degree program and spend a great deal of time at work utilizing their specialized training. For this reason, we precede our study of design by reviewing the characteristic features of the various branches of engineering.

Aeronautical Engineer

Aeronautical (or aerospace) engineers use their knowledge of aerodynamics, fluid mechanics, structures, guidance and control systems, heat transfer, and hydraulics to design and build everything from airplanes, rockets, and space shuttles to high-speed bullet trains and helium-filled dirigibles. Since the days of the Wright brothers, aeronautical engineers, working in teams with scientists and other types of engineers, have made possible human flight and space exploration. Aeronautical engineers find employment in many industries, but typically work for big companies on large-scale projects involving many engineers. Some of the more noticeable accomplishments of the aerospace industry have included the Apollo moon landings, the NASA Space Shuttle, deep space exploration, and the jumbo jet. The space station of Figure 1.1, for example, will be built by teams of aeronautical and other engineers.

Agricultural Engineer

Agricultural engineers apply the principles of hydrology, soil mechanics, fluid mechanics, heat transfer, combustion, optimization theory, statistics, climatology, chemistry, and

Figure 1.1. Space stations like the one depicted in this artist's view will be built by teams of aerospace and aeronautical engineers working together with other types of engineers. (*Photo courtesy of NASA.*)

biology to the production of food on a large scale. This discipline is popular at colleges and universities located in heavily agricultural areas. Feeding the world's ever-growing population is one of the most formidable challenges of the 21st century. Agricultural engineers will play an important role in this endeavor by applying technology and engineering know-how to improve crop yields, increase food output, and develop cost-effective and environmentally sound farming methods. Agricultural engineers work with ecologists and natural scientists to understand the impact of human agriculture on the earth's ecosystem.

Biomedical Engineer

The biomedical engineer (or bioengineer) applies modern engineering methods and technology to solve problems in medicine and human health. The biomedical engineer uses quantitative methods and works closely with physicians and biologists to obtain a better understanding of the human body. Engineering skills are combined with knowledge of biology, physiology, and chemistry to produce medical instrumentation, prosthetics, appliances, implants, and neuromuscular diagnostics. Biomedical engineers have participated in designing many devices that have helped improve medical care over the past several decades. Many biomedical engineers enter medical school upon graduation, but others go on to graduate school or seek employment in any of a number of health- or medical-related industries. One newly emerging branch of biomedical engineering, called molecular engineering, examines the fundamental functions of cells and organisms from an engineering point of view. The science of cloning, for example, has been made possible in part by the work of molecular engineers. Many of the secrets

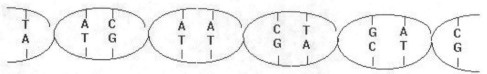

Figure 1.2. Genetics is one of the new frontiers of biomedical engineering.

of future medicine lie at the genetic level, and the biomedical engineer will help lead the way to new medical discoveries.

Chemical Engineer

A chemical engineer applies the principles of chemistry to the design of manufacturing and production systems. Whenever a chemical reaction or process must be brought from the laboratory to manufacturing on a large scale, a chemical engineer is needed to design the transport mechanisms, mixing chambers, measuring devices, and reaction vessels that allow the reaction to proceed on a large scale in a cost effective way. Chemical engineers are employed in many industries, including petroleum, petrochemicals, plastics, cosmetics, electronics, food, and pharmaceuticals. Their skills are needed anywhere that a manufacturing process involves organic or inorganic chemical reactions on a production scale.

Civil Engineer

The civil engineer is concerned with the design and construction of roads, bridges, buildings, airports, and other large structures, such as water treatment plants, aquifers, transportation systems, and the Hoover Dam shown in Figure 1.3. Designing on such a large scale requires knowledge of structural, fluid, and soil mechanics, strength of materials, and construction practices. Over the next few decades, civil engineers will play a vital role in revitalizing the nation's aging infrastructure and in dealing with environmental issues, such as water resources, air quality, global warming, and refuse disposal. More than any other discipline, civil engineering has the unique handicap of having to rely heavily on scale models, calculations, computer modeling, and past experience to determine the performance of designed structures. The civil engineer works closely with construction personnel and may spend much time at job sites reviewing the progress of construction tasks. Civil engineers often are employed in the public sector, but may also find work in large construction companies and private development firms.

Computer Engineer

Computer engineering encompasses the broad categories of hardware, software, and digital communication. A computer engineer applies the basic principles of engineering

Figure 1.3. The Hoover Dam at the Nevada-Arizona border was designed by civil engineers.

Figure 1.4. Civil engineers will be responsible for revitalizing the nation's infrastructure in the 21st century.

and computer science to the design of computers, networks, peripheral devices, and software systems. The computer engineer has the responsibility of designing and building the interconnections between computers and their components, including desktop PCs, local area networks (LANs), and Internet servers. For example, a computer engineer might combine microprocessors, memory chips, disk drives, CD drives, display devices, modems, and drivers to produce computer systems. Graphical user interfaces, embedded computer systems, fault-tolerant computers, software systems, and assembly language programming are also the responsibility of the computer engineer. Computer scientists, who traditionally are more mathematically oriented than computer engineers, also have become involved in the writing of computer software over the past decade. Unlike the computer scientist, however, the computer engineer is fluent in both the hardware and software aspects of modern computer systems. Examples of systems in which both hardware and software share equally important roles include telephone and communications systems, process control, automation systems, control systems, management information systems, microcomputer-controlled appliances, and medical instrumentation. Some of the more notable accomplishments of the computer industry include the invention of the microprocessor (Intel, 1982), the explosion of personal computing that began with the first desktop PC (IBM, 1984), and the advances in data communication networks that began with the U.S. Department of Defense Arpanet and grew into the Internet and World Wide Web.

Electrical Engineer

Electrical engineering encompasses a wide range of disciplines linked by a single common thread: the use and control of electric, electronic, or electromagnetic energy. Whether on a large or small scale, electrical engineers are responsible for numerous

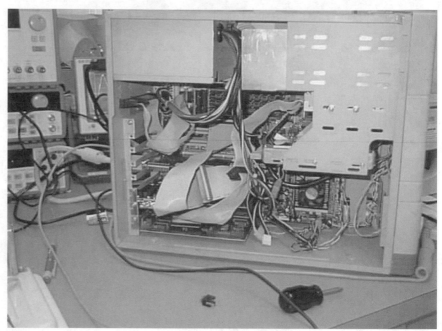

Figure 1.5.   Computer engineers design the hardware and software for today's computer systems.

areas of technology, including microelectronics, speech recognition, data communication, radio, television, lasers, fiber optics, video, audio, computer networks, electric power systems, and alternative energy sources, such as solar and wind power. The electrical engineer also designs transportation systems based on electric power, including mass transit and electric cars. The typical electrical engineer has a strong background in the physical sciences, mathematics, and computational methods, as well as knowledge of circuits and electronics, semiconductor devices, analog and digital signal processing, digital systems, electromagnetics, and control systems. The electrical engineer also is fluent in many areas of computer engineering. Some of the more recent accomplishments that have involved electrical engineers include the microelectronic revolution (microprocessors, large-scale integration on a chip), wireless communications (cellular phones, pagers, and data links), photonics (lightwave technology, lasers, and fiber-optic communication), and micro-electromechanical systems (**MEMS**).

Industrial Engineer

Industrial engineering (sometimes called manufacturing engineering) is concerned with the design of devices and systems for mass production and materials processing. Industrial engineers have the unique challenge of incorporating the latest technological advances in computing and machinery into production and manufacturing facilities. The industrial engineer is intimate with all aspects of the corporate environment, because much of what motivates the field of industrial engineering is the need to maximize output while minimizing cost. Skills required for this discipline include knowledge of product development, materials processing, optimization, queuing theory, production techniques, and engineering economy. Industrial engineers also become fluent in the techniques of computer-aided design (CAD) and computer-aided manufacturing (CAM). One of the more recent areas to emerge as the province of the industrial engineer is the use of robotics in manufacturing. Building, moving, and controlling robots

Figure 1.6.   The laser, first invented around 1960, has become an indispensable tool for the electrical engineer.

requires knowledge of aspects of mechanical, electrical, and computer engineering. Most programs in industrial engineering include courses in these other areas.

Mechanical Engineer

The mechanical engineer is responsible for designing and building physical structures. Devices that involve mechanical motion, such as automobiles, bicycles, engines, disk drives, keyboards, fluid valves, jet engine turbines, power plants, and flight structures, are all made by mechanical engineers. Mechanical engineers are fluent in the topics of statics, dynamics, strength of materials, structural and solid mechanics, fluid mechanics, thermodynamics, heat transfer, and energy conversion. They apply these principals to a wide variety of engineering problems, including acoustics, precision machining, environmental engineering, water resources, combustion, power sources, robotics, transportation, and manufacturing. Because of their broad educational background, mechanical engineers interface easily with most any other type of engineer.

One of the newest areas of study is the emerging field of micro-electromechanical systems (MEMS) in which tiny microscopic machines are fabricated on wafers of silicon and other materials. Figure 1.7 for example shows a tiny micromotor built on a silicon chip. MEMS has the potential to do for mechanics what the integrated circuit did for electronics, namely permit large-scale integration of small structures on single silicon "chips." Mechanical engineers work closely with electrical engineers in this exciting new discipline

Mechatronics Engineer

As its name implies, the field of mechatronics involves the fusion of mechanical engineering, electronics, and computing toward the design of products and manufacturing

Figure 1.7. Tiny MEMS silicon micromotor measuring 100 micrometers in diameter. (*Photo courtesy of MCNC.*)

systems. Engineers who work in this emerging field require cross-disciplinary training that can best be approached by majoring in either mechanical or electrical engineering and acquiring skills in the other needed disciplines through extra courses or technical electives. Mechatronics engineers are responsible for the innovation, design, and development of machines and systems that can reduce production costs, reduce plant maintenance costs, improve product flexibility, and increase production performance. The typical mechatronics engineer solves design problems for which solely mechanical or electrical solutions are not possible. Sensing and actuation are important elements of mechatronics.

Naval Engineer

The naval engineer designs ships, submarines, barges, and other seagoing vessels, and is also involved in the design of oil platforms, shipping docks, seaports, and coastal navigation facilities. Naval engineers are fluent in many of the subjects studied by mechanical engineers, including fluid mechanics, materials, structures, statics, dynamics, water propulsion, and heat transfer. In addition, naval engineers take courses on the design of ships and the history of sea travel. Many naval engineers are employed by the armed forces, but some work for companies that design and build large ships.

Petroleum Engineer

Over seventy percent of the world's current energy needs are derived from petroleum products, and this situation is unlikely to change for at least the next half century. The principal challenge of the petroleum engineer is to produce oil, gas, and other energy forms from the earth's natural resources. In order to harvest these resources in an economical and environmentally safe way, the petroleum engineer must have a wide base of knowledge that includes mathematics, physics, geology, and chemistry, as well as aspects of most all other engineering disciplines. Elements of mechanical, chemical, electrical, civil, and industrial engineering are found in most programs in petroleum engineering. Also, because computers are used with ever-increasing frequency in geological exploration, oilfield production, and drilling operations, computer engineering has become an important specialty within petroleum engineering. Did you know that many of the world's supercomputers are owned by petroleum companies?

In addition to conventional oil and gas recovery, petroleum engineers apply new technology to the enhanced recovery of hydrocarbons from oil shale, tar sands, offshore oil deposits, and fields of natural gas. They also design new techniques for recovering residual ground oil that has been left by traditional oil-pumping methods. Examples include the use of underground combustion, steam injection, and chemical water treatment to release oil trapped in the pores of rock. These techniques also are likely to be used in the future for other geological operations, such as uranium leaching, geothermal energy production, and coal gasification. Petroleum engineers also work in the related areas of pollution reduction, underground waste disposal, and hydrology. Lastly, because many petroleum companies operate on a worldwide scale, the petroleum engineer has the opportunity to work in numerous foreign countries.

Systems Engineer

The traditional systems engineer is concerned with the design and implementation of large-scale systems. Programs of study in this broad field include courses in applied mathematics, computer simulation, software, electronics, communications, and automatic control. Systems engineers are at home working with many different types of engineers, including electrical, mechanical, and computer engineers, and find work in data processing, power generation and transmission, communications, aerospace, and public utilities. In some circles, especially the computer industry, the designation "systems engineer" has come to mean someone who specifically deals with large-scale computer and software systems.

1.2 ENGINEERING PROFESSIONAL ORGANIZATIONS

Most of the branches of engineering are represented by professional societies that bond together members of similar background, training, and professional expertise. Some societies operate on a world-wide scale, and most publish one or more journals in which engineers publish papers and articles of interest to the field. Each organization offers its members technical and informational services, and in some cases other professional services, such as life insurance, job networks, advertising, e-mail access, and Web page hosting. All provide student membership at a discount, and student chapters at colleges and universities are common. This section provides information about some of the principal professional organizations and the technical publications they produce. Each society has an official Web page from which you can obtain additional information. The text provided here has been taken from each organization's Web page.

Aeronautical and Aerospace Engineering

American Institute of Aeronautics and Astronautics (source: http://www.aiaa.org)

> "The nonprofit American Institute of Aeronautics and Astronautics (AIAA) is the principal society and voice serving the aerospace profession. Its primary purpose is to advance the arts, sciences, and technology of aeronautics and astronautics and to foster and promote the professionalism of those engaged in these pursuits. Although founded and based in the United States, AIAA is a global organization with nearly 30,000 individual professional members, over 50 corporate members, thousands of customers worldwide, and an active international outreach. AIAA is the U.S. representative on the International Astronautical Federation and the International Council on the Aeronautical Sciences."

Key Publications   *Aerospace America, AIAA Bulletin, Aerospace Database, Student Journal*

Biomedical Engineering

Biomedical Engineering Society (source: http://www.mecca.org/bme/bmes/bmeshome. html)

> "The Biomedical Engineering Society (BMES) is an interdisciplinary society established on February 1, 1968 in response to a manifest need to provide a society that gave equal status to representatives of both biomedical and engineering interests. As stated in the Articles of Incorporation, the purpose of the Society is: 'To promote the increase of biomedical engineering knowledge and its utilization.' Today, the society represents over 1000 professionals and over 1000 student members (undergraduate and graduate). There are 34 BMES student chapters and about two-thirds of the ABET-accredited bioengineering/ biomedical engineering undergraduate programs have BMES student chapters."

Key Publication   "The BMES scientific journal, *The Annals of Biomedical Engineering* (ABME), is published bimonthly under contract with the American Institute of Physics and presents original research in the following areas: tissue and cellular engineering and biotechnology; biomaterials and biological interfaces; biological signal processing and instrumentation; biomechanics, rheology, and molecular motion; dynamical, regulatory, and integrative biology; transport phenomena, systems analysis and electrophysiology; and imaging."

Chemical Engineering

American Institute of Chemical Engineers (source: http://www.aiche.org)

> "Founded in 1908, the American Institute of Chemical Engineers (AIChE) is a nonprofit organization providing leadership to the chemical engineering profession. Representing 58,000 members in industry, academia, and government, AIChE provides forums to advance the theory and practice of the profession, upholds high professional standards and ethics, and supports excellence in education. Institute members range from undergraduate students and entry-level engineers to chief executive officers of major corporations."

Key Publications   "*Chemical Engineering Progress* (CEP), AIChE's flagship monthly magazine, reports on recent advances in chemical process and related industries. CEP also annually publishes a comprehensive software directory of programs of interest to chemical engineers."

Civil Engineering

American Society of Civil Engineers (source: http://www.asce.org)

> "Founded in 1852, ASCE represents more than 120,000 civil engineers worldwide and is America's oldest national engineering society. ASCE advances professional knowledge and improves the practice of civil engineering as the lead professional organization serving civil engineers and those in related disciplines. It serves as the focal point for development and transfer of research results, and technical, policy and managerial information and is the catalyst for effective and efficient service through cooperation with other engineering and related organizations."

Key Publications   *ASCE News, Civil Engineering Magazine*

Computer Engineering

Association for Computing Machinery (source: http://www.acm.org)

"The Association for Computing Machinery (ACM) is an international scientific and educational organization dedicated to advancing the arts, sciences, and applications of information technology. With a world-wide membership of 80,000, ACM functions as a locus for computing professionals and students working in the various fields of Information Technology.

"ACM publishes, distributes, and archives original research and firsthand perspectives from the world's leading thinkers in computing and information technologies. ACM offers over two dozen publications that help computing professionals negotiate the strategic challenges and operating problems of the day. The ACM Press Books program covers a broad spectrum of interests in computer science and engineering.

"*Communications of the ACM* keeps information technology professionals up to date with articles spanning the full spectrum of information technologies in all fields of interest including object oriented technology, multimedia, Internetworking, and hypermedia to name just a few. *Communications* also carries case studies, practitioner-oriented articles, and regular columns, the ACM Forum, and technical correspondence. The monthly magazine is distributed to all ACM members and has recently made its appearance at selected newsstands.

"The *netWorker* magazine is the only publication to fully examine network computing and all its applications and implications. The netWorker's editorial staff receives ongoing contributions from an expert advisory board who offer insights, suggestions, and articles."

Key Publications   ACM Transactions Journals: *Computer-Human Interaction, Computer Systems, Database Systems, Design Automation for Electronic Systems, Graphics, Information Systems, Mathematical Software, Modeling and Computer Simulation, Networking, Programming Languages and Systems, Software Engineering and Methodology.*

Electrical Engineering

Institute of Electrical and Electronic Engineers (source: http://www.ieee.org)

"The Institute of Electrical and Electronics Engineers (IEEE) is one of the world's largest technical professional societies. Founded in 1884 by a handful of practitioners of the new electrical engineering discipline, today's Institute is comprised of more than 320,000 members who conduct and participate in its activities in approximately 150 countries. The men and women of the IEEE are the technical and scientific professionals making the revolutionary engineering advances which are reshaping our world today.

"The technical objectives of the IEEE focus on advancing the theory and practice of electrical, electronics and computer engineering and computer science. To realize these objectives, the IEEE sponsors technical conferences, symposia, and local meetings worldwide. It publishes nearly 25% of the world's technical papers in electrical, electronics, and computer engineering, and provides educational programs to keep its members' knowledge and expertise state-of-the-art."

Key Publications   *IEEE Spectrum, Proceedings of the IEEE,* plus over 40 specialized *IEEE Transactions* from its various societies. A partial list of IEEE technical societies includes the following groups: Aerospace and Electronic Systems, Antennas and Propagation, Circuits and Systems, Communications, Computer, Control Systems, Electron Devices, Industry Applications, Information Theory, Lasers & Electro-Optics Magnetics, Microwave Theory and Techniques, Neural Networks Council, Power Engineering, Robotics & Automation, Signal Processing, and Solid-State Circuits.

Industrial and Manufacturing Engineering

Institute of Industrial Engineers (source: http://www.iienet.org)

> "Founded in 1948, the Institute of Industrial Engineers is the society dedicated to serving the professional needs of industrial engineers and all individuals involved with improving quality and productivity. Its 24,000 members throughout North America and more than 80 countries stay on the cutting edge of their profession through IIE's life-long-learning approach, as reflected in the organization's educational opportunities, publications, and networking opportunities. Members also gain valuable leadership experience and enjoy peer recognition through numerous volunteer opportunities."

Key Publications   *IIE Solutions, Industrial Management, IIE Transactions, The Engineering Economist, Student IE, Journal of the Society for Health Systems.*

Mechanical Engineering

American Society of Mechanical Engineers (source: http://www.asme.org)

"With 125,000 mechanical engineers as members, ASME International:

- Offers quality programs and activities in mechanical engineering enabling its practitioners to contribute to the well-being of humankind.
- Conducts conferences, exhibits and regular meetings of local chapters to keep practicing mechanical engineers up to date on new technology.
- Publishes 19 technical journals, plus numerous books, technical reports and magazines on mechanical engineering.
- Facilitates the development and application of technology in areas of interest to ASME members and the mechanical engineering profession.
- Maintains and distributes 600 codes and standards for the design, manufacturing, and installation of mechanical devices.
- Provides short courses on current technical developments.
- Advises federal and state government on technology-related public policies.
- Manages ASMENET on the World Wide Web (http://www.asme.org).
- Disseminates information about mechanical engineering and technology to elementary and high school students and to the general public."

Key Publications   "ASME International conducts one of the largest technical publishing programs in the world. Its general-interest publications include:

Mechanical Engineering, the official ASME International monthly magazine since 1919, which explores the design, development and application of the wide-ranging technologies of interest to mechanical engineers.

ASME News, which reports the significant events of the Society and the accomplishments of members.

The monthly *Applied Mechanics Reviews,* a critical review journal, covers rational mechanics, thermal sciences, mechanics of solids, automatic control, and mechanics of fluids. A bimonthly publication is *Heat Transfer: Recent Contents.*

"ASME International also publishes 19 quarterly transactions journals that carry technical papers selected for their archival and technical value. These journals cover all aspects of mechanical engineering, from applied mechanics to vibration and acoustics."

PROFESSIONAL SUCCESS: CHOOSING A FIELD OF ENGINEERING

If you are a first-year student of engineering, you may already have decided upon a major field. After taking a number of required courses, however, you may not be sure if you've chosen the right type of engineering. Conversely, you may have entered school without committing yourself to a particular branch of engineering. If you find yourself in either of these situations, you're probably wondering how one goes about choosing an engineering specialty.

One way to find out more about the different branches of engineering is to attend talks and seminars hosted by departments in your college or university. Such talks usually are aimed at graduate students and faculty, so expect much of the material to be over your head. But by simply exposing yourself to different technical talks, you can get a feeling for the various branches of engineering and help yourself find one that most closely matches your skills and interests.

Some schools host workshops in career advising. Be sure to attend one. You'll get a chance to talk with experts in career planning and job placement. Many college campuses host student chapters of professional organizations. These groups often organize tours of engineering companies. Attending such a tour can orient you toward the activities of a particular branch of engineering and provide you with a sampling of what life as an engineer will be like.

One of the most valuable resources for career advise is your own college faculty. Professors love to talk about their work and careers. Get advice about which major is right for you. Invite a professor to your dormitory or living unit to speak to students about choosing an engineering career. Speak to your department about hosting a career night in which a panel of professors answers questions about jobs in engineering. Learn to make use of all available resources for help in choosing a major subject.

1.3 THE ENGINEER: CENTRAL TO PROJECT MANAGEMENT

When we think of the word "design," we may imagine an individual engineer sitting alone in a cubicle at a computer terminal, or perhaps in a workshop, crafting some marvelous piece of technical wizardry. As a student, you may be eager pursue this notion and become a sole entrepreneur who single-handedly changes the face of technology. You might ask, "Why do I have to take all these *other* courses? Why can't I just take courses that are of interest to me or important to my career?" The answer to these questions lies in the multidisciplinary nature of engineering. At times, an engineer does work alone, but most of the time, engineers must interface with individuals who come from a wide variety of educational backgrounds. Engineering projects can be complex undertakings that require the coordination of many people of different skills and personality traits. It's been said that a good engineer acts as the glue that ties a project together, because he or she has learned to communicate with specialists from many different fields. An engineer must learn the languages of physicists, mathematicians, chemists, managers, fabricators, technicians, lawyers, marketing staff, and secretaries. Learning to communicate with each of these very different types of people requires that the engineer have a broad education and the ability to apply principles from many different fields to the design process.

The Well Rounded Engineer

To help illustrate the breadth of communication skills required of an engineer, imagine that you work for the fictitious company depicted in Figure 1.8. Each person shown in the outer circle represents a different professional expertise represented by a famous person of appropriate background. Notice that you, the design engineer, lie in the center of the organizational circle. Other engineers on your design team may join you in the center, but each of you can easily communicate with any one specialist in the outer ring, because, as an engineer, you've taken courses or have been exposed to each of their various disciplines. This unique feature of your educational background enables to you communicate with anyone in the professional circle and positions you as an individual most able to act as central coordinator.

The Physicist (*e.g., Albert Einstein, best known for his theory of relativity*)　　The physicist of the company is responsible for understanding the basic physical principles that underlie the company's product line. He spends his time in the laboratory exploring new

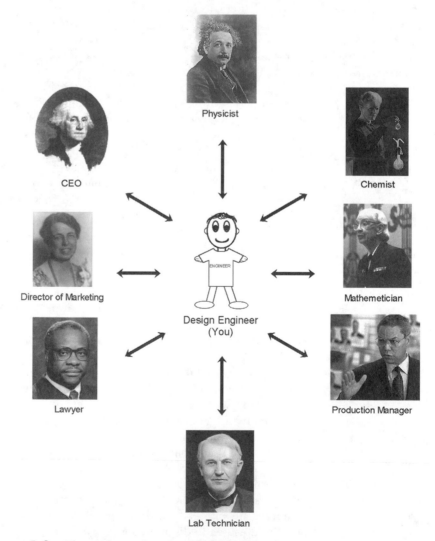

Figure 1.8.　The professional circle with the design engineer at its center.

materials, analyzing their interactions with heat, light, and electromagnetic radiation. He may discover a previously unknown quantum interaction that will lead to a new semiconductor device or perhaps explore the potential for using superconductors in the company's product line. Or, he may simply perform the physical analysis of a concept for a new microaccelerometer. Because you have taken two or more semesters of basic physics, you understand something about mechanics, thermodynamics, and electromagnetics and can communicate with the physicist about how these basic discoveries might translate into practical devices of interest to the company.

The Chemist (*e.g., Marie Curie, who discovered radium*)   The chemist performs analyses on substances that are used in producing company products. She ensures that raw materials used for manufacturing meet purity specifications so that quality control can be maintained. In her laboratory, she may direct a team of technicians in experiments seeking improved materials that are stronger and more durable than those now being used. She may perform research on complex organic compounds or perhaps work on molecular-based nanotechnology. As an engineer, you've taken one or more courses in chemistry and can speak her language. You understand such concepts as reaction ratios, chemical equilibrium, molarity, reduction and oxidation, acids and bases, and electrochemical potential. Perhaps you are a software engineer writing program code for a chemical analysis instrument. Maybe you are a manufacturing engineer charged with translating a chemical reaction into a manufactured product. Whatever your role, you are an individual very well suited to bringing the contributions of the chemist to the design process.

The Mathematician (*e.g., Grace Hopper, former Navy Admiral, mathematician, and computer specialist who coined the term "bug"*)   The mathematician of the company, who might also be a computer scientist, worries about such things as modeling, statistics, databases, and forecasting. She may be involved in an intriguing new database algorithm or mathematical method for modeling an engineering system. Perhaps she uses mathematics to analyze the company's production line or to forecast trends in marketing. You converse easily with the mathematician, because you have taken numerous math courses as part of your engineering program. Although your emphasis has been on applied rather than pure mathematics, you're familiar with calculus, differential equations, linear algebra, statistics, probability, and complex variables. You can easily translate the concepts of mathematics to problems in engineering design.

The Production Manager (*e.g., Colin Powell, former U.S. Army general, military planner, and co-architect of Operation Desert Storm*)   Like the army general in top command, the production manager has the responsibility of mobilizing materials, supplies, and personnel to manufacture company products. The production manager may worry about such things as job scheduling, quality control, materials allocation, quality assurance testing, and yield. As the engineer who designs products, you work closely with the production manager to make sure that your design approach is compatible with the company's manufacturing capabilities. Your training as an engineer and your exposure to machining, welding, circuit fabrication, and automation have given you the ability to understand the job of the production manager and have provided you with the vocabulary needed to communicate with him.

The Lab Technician (*e.g., Thomas Edison, famous tinkerer and experimenter, best known for inventing the incandescent light bulb and phonograph*)   The lab technician an indispensable member of the design team. A habitual tinkerer and experimenter, the lab technician helps bring your design project to fruition. He is adept at using tools and has much knowledge about the practical aspects of engineering. The lab technician is masterful at fabricating prototypes and is likely to be the individual who sets up and tests them. The typical lab technician has a degree in engineering technology closely related to a field of engineering. You and the lab technician have taken many courses in common, although your courses probably have included more formal theory and mathematics than his. You can communicate easily with the lab technician and include him at each phase of your design project.

The Lawyer (*e.g., Clarence Thomas, lawyer and Supreme Court Justice*)   The lawyer worries about the legal aspects of the company's products. Should we apply for a patent on the XYZ widget? Are we exposing ourselves to a liability suit if we market a substandard product? Is our new deal with Apex Co. fair to both companies from a legal perspective? To help the lawyer answer these and other questions, you must be able to communicate with him or her and share your engineering knowledge. The logical thought that forms the basis of the law is similar to the methods that you've used to solve countless engineering problems. As an engineer, you easily engage in discourse with the lawyer and can apply his or her legal concerns about safety, ethics, and liability to the design process.

The Director of Marketing (*e.g., Eleanor Roosevelt, former First Lady of the United States*)   The director of marketing is a master of imagery and style. She has the job of selling the company's products to the public and convincing people that your products are better than those of your competitors. The marketing manager has excellent communication skills, some knowledge of economics, and an understanding of what makes people want to buy. You interface easily with the marketing manager because you've dealt with all aspects of design as part of your training as an engineer. Through this training, you have focused not only on technical issues, but on such things as product appearance, the human-machine interface, durability, safety, and ease of use. Your familiarity with these important issues has prepared you to help the director of marketing understand your product and how it works. Similarly, you are able to respond to her concerns regarding what the public needs from the product you design.

The President/Chief Executive Officer (*e.g., George Washington, first president of the United States*)   The CEO of the company probably has an MBA (Master's of Business Administration) or higher degree and a long history working at corporate financial affairs. The CEO worries about the economy and what future markets the company may pursue or whether to open a new plant in a foreign country. It's the CEO who determines how your current project will be financed, and he needs to be kept up to date about its progress. The CEO also may ask you to assess the feasibility of a new technology or product concept. As an engineer, you have no difficulty conversing with the CEO, because the economic principles of profit and loss, cost derivatives, statistics, and forecasting are closely tied to concepts you learned in courses on calculus, statistics, and economics. You've probably learned to use spreadsheets in one or more engineering classes and have no trouble interpreting or providing the information that is part of the CEO's world. Likewise, your training as an engineer prepares you to communicate with the CEO about the impact of your design project on the economic health of the company.

1.4 ENGINEERING: A SET OF SKILLS

To be successful at design, an engineer must acquire many skills. An engineer must have technical, theoretical, and practical skills and must be good at organization, communication, and documentation. Three especially important traits that form the foundation of an engineer's competency are *knowledge, experience,* and *intuition.* These talents do not form an exhaustive set, but they help form the hallmark of a competent, well-rounded engineer.

Knowledge

Knowledge consists of the body of facts, scientific principles, and mathematical tools that an engineer uses to form strategies, analyze systems, predict results, or seek a deeper understanding of how something works. An engineer acquires knowledge by studying many subjects. The natural sciences, such as physics, chemistry, and biology, help an engineer understand the physical world. Mathematics provides a universal technical language that bridges different disciplines, spoken languages, and cultural boundaries. Each field of engineering is based on a traditional body of knowledge, but an engineer in one field also learns subjects from other fields. Areas of knowledge that are common to most engineers, regardless of discipline, include mechanics, circuits, materials science, and computer programming. As a student of engineering, you may ask why you are required to take subjects that seem irrelevant to your career aspirations. Any experienced engineer will tell you the answer: Engineers work in a multidisciplinary world where basic knowledge of many different subject areas is a necessity. Mechanical and computer engineers use electrical circuits. Electrical engineers build physical structures and use computers. Aeronautical engineers rely on software systems. Software engineers design airplane controls. Understanding the field of another engineer is critical to cross-disciplinary communication and design proficiency.

Although formal education is an important part of any engineer's training, the prudent engineer also acquires knowledge through on-the-job training and a lifetime of study and exploration. Tinkering, fixing, experimenting, and taking things apart to see how they work are important sources of engineering knowledge. As a young person, did you disassemble your toys, put model kits together, write your own software, create Web pages, or play with building sets, hammers, nails, radios, bicycles, or computers? Without knowing it, you began the path toward acquiring engineering knowledge. The professional engineer engages in these same practices. By becoming involved in all aspects of a design project, by keeping up to date with the latest technology, taking professional development courses, and solving real world problems, the practicing engineer remains current and competent.

Experience

Experience refers to the body of procedures, methods, techniques, and rules of thumb that an engineer uses to solve problems. Accumulating experience is just as important to an engineer's career as acquiring knowledge. As a student, you will have several opportunities to gain engineering experience. Cooperative assignments, assistantships in labs, capstone design projects, summer jobs, and research work in a professor's laboratory provide important sources of engineering experience. On-the-job training is also a good way to gain valuable professional experience. Many engineering companies recognize this need and provide entry-level engineers with initial training as a way of infusing additional experience. Developing experience requires "seasoning," the process by which a novice engineer gradually learns the rules of thumb and "tricks of the trade" from other, more experienced engineers. Company lore about methods, procedures, and a history of how things are done is often passed orally, from one generation of engineers to the next, and a new engineer learns this information by working with other engineers. The history of what *hasn't* worked in the past is also a key part of this lore.

An engineer also gains valuable experience by enduring design failure. When the first attempt at a design fails in the testing phase, the wise engineer views it as a learning experience and uses the information to make needed changes and alterations. Experience is acquired by testing prototypes, studying failures, and observing the results of design decisions.

Engineers also must consider the issues of reliability, cost, manufacturability, ergonomics, and marketability when making design decisions. Only by confronting these real-world constraints in addition to technical concerns can an engineer gain design experience.

Intuition

Intuition is a characteristic normally associated with baseball players, fortune tellers, and weather forecasters. Intuition is also an essential element of engineering. It refers to an engineer's basic instinct about what will or will not work when trying to solve engineering problems. Although intuition can never replace the careful planning, analysis, and testing that are part of good design, and at times can act to suppress innovation, it also can help an engineer decide which approach to follow when faced with choices and no obvious answer. A feeling for what will work and what will not work, based solely on extensive experience, can save time by helping an engineer choose the path that will eventually lead to success rather than failure. Intuition helps an engineer predict whether a design concept will work before it's actually built. When intuition is at work, you might hear an engineer use such phrases as, "it seems reasonable," "that looks about right," or "oh, about this much."

Intuition is a direct byproduct of design experience and is acquired only through practice, practice, and more practice. In the information age, where much of engineering focuses on the computer, engineers are tempted to solve everything by simulation and computer modeling. It's easy to forget that engineering systems ultimately must interact with people and obey the idiosyncrasies of the real, physical world. Developing intuition and becoming comfortable with that world is an important part of your engineering education. How many times have you opened your computer to install new components or to adjust hardware settings? Have you tinkered with the family car or your bicycle? Have you built something from a kit or raw materials? Each of these tasks helps you acquire intuition. Observing the way in which the boards of a computer have been laid out by other engineers will acquaint you with the techniques of hardware design. Adjusting the gear and brake settings of your bicycle will help you to understand the tradeoffs inherent to good mechanical design, such as the conflict between strength and durability versus lightweight construction. Becoming knowledgeable in the use of tools will help you better understand the impact of your design decisions on manufacturing. Repetition, testing, careful attention to detail, working with more experienced engineers, and dedication to your discipline are the keys to developing design intuition. Design intuition is best acquired by "doing design" and by playing with real things.

PROFESSIONAL SUCCESS: HOW TO GAIN EXPERIENCE AS A STUDENT

Chapter 1 stresses the importance of experience in the life of an engineer. You can begin to accumulate design experience even while you are a student. A cooperative education program, if your school has one, is an excellent way to gain experience as a engineer. The typical program places you as an intern in an engineering company for six to twelve months. You'll typically be assigned to a senior engineer to assist in computer aided design work, software development, product prototyping, testing, laboratory evaluation, or other tasks. You'll get to see how the company works, and the company will be able to evaluate you as a possible future hire. In addition, you'll be paid for the time you spend at the company. You be able to acquire real engineering experience and will not have to pay tuition at the same time.

Students also can gain valuable experience by working in research labs at school. Your faculty's research interests are probably listed on the back pages of your college catalog and on your department's Web site. Learn about the research activities of a professor you've enjoyed having in class. Don't be afraid to simply ask if he or she needs help in the lab. Many professors receive industry or government funding for their research, so you may even be paid a small stipend or hourly wage for your time. You'll be assigned such tasks as constructing circuits or test fixtures, writing software, taking data, preparing and testing samples, and assisting graduate students.

KEY TERMS

Engineering	Profession	Knowledge
Experience	Intuition	

2

What Is Design?

de·sign n: the arrangement of parts, details, form, color, etc. so as to produce an artistic or skillful invention*

Throughout history, the engineer has made useful things from basic raw materials. Even in ancient times, the industrious and inventiveness of the engineer were highly regarded by society. The work of the engineer has left its mark on every era of civilization, from the great pyramids of Egypt, the Roman aqueducts, the rope bridges of Nepal, the Great Wall of China, and the ancient Mayan temples to the Eiffel tower, the Golden Gate Bridge, and the transatlantic cable. From a fundamental perspective, design can be defined as any activity that results in the synthesis of something that meets a need. A refrigerator keeps food cold, a bicycle provides transportation, a keyboard sends data to a computer, and a muffler silences a noisy automobile engine. Although design is practiced every day by many creative people, the notion of "design" in the context of engineering usually implies that knowledge is combined with specialized skills to create a device, machine, circuit, building, mechanism, structure, software program, manufacturing process, or other system that meets a set of desired specifications. In this latter usage, the word "design" answers the simple question, "What do engineers do?"

*Webster's New World Dictionary of American English. V. Neufeldt, ed. New York: Prentice Hall, 1994.

SECTIONS

- 2.1 Use of the Word Design
- 2.2 The Difference Between Analysis, Reproduction, and Design
- 2.3 Good Design Versus Bad Design
- 2.4 The Design Cycle
- 2.5 A Design Example

OBJECTIVES

In this chapter, you will learn about:

- The engineering design process.
- The difference between good design and bad design.
- The design cycle.
- The Peak Performance Design Competition.

2.1 USE OF THE WORD DESIGN

Throughout this book, the word "design" will be used in several ways. It may be used as a verb, as in, "Design a widget that can open a soda can automatically," or perhaps as a noun that defines the creation process itself, as in, "Learning design is an important part of engineering education." Sometimes, design will be used as a noun that describes the end result of the process, as in "The design was successful and met customer specifications." At other times, the word will be used as an adjective, as in, "This book will help you learn the design process."

Sometimes, an alternative word will be needed to describe the end product of a design effort. For this purpose, the word "product" may be used in a generic sense, even if the thing being designed is not a product for sale. Similarly, the word "device" may be used to describe the results of a design effort, even if the entity is not a physical apparatus. Thus the words product and device will refer not only to tangible objects, but also to large structures, systems, procedures, and software.

2.2 THE DIFFERENCE BETWEEN ANALYSIS, REPRODUCTION, AND DESIGN

Students of engineering often are confused by the distinction between analysis, reproduction, and design. In science classes, students are asked to find answers to problems, complete laboratory exercises, and perform calculations. In engineering classes, instructors instead may stress the importance of design. The difference between analysis and design can be defined in the following way: If only one answer to the problem exists, and finding it merely involves putting together the pieces of the puzzle, then the activity is probably analysis. For example, processing data and using it to test a theory, is analysis. On the other hand, if more than one solution exists, and if deciding upon a suitable path demands creativity, choice taking, testing, iteration, and evaluation, the activity is most certainly design. Design can include analysis, but it also must involve at least one of these other elements.

As an example of the distinction between analysis and design, consider the weather station shown in Figure 2.1. This remote-controlled buoy is anchored about ten miles off the coast of Massachusetts and is maintained for the U.S. National Oceanic and Atmospheric Administration. It provides 24-hour data to mariners, the Coast Guard, and weather forecasters. Processing the data stream from this buoy, deciding which parts to post on the Internet*, and using the information to forecast the weather are examples of analysis. Deciding *how* to build the buoy so that it meets the needs of its end users is an example of design.

Another example that illustrates the difference between analysis and design can be found in the medieval catapult of Figure 2.2. Determining the projectile's x-y trajectory is an analysis problem that involves the solution of the following equations:

Newton's law $\mathbf{F} = m\mathbf{a}$:

$$d^2x/dt^2 = 0$$
$$d^2y/dt^2 = mg$$

Initial conditions at $t = 0$:

$$dx/dt = v_x = V_0 \cos \theta$$
$$dy/dt = v_y = V_0 \sin \theta$$

*http://www.ndbc.gov

Figure 2.1. NOAA buoy off the coast of Massachusetts provides information about local sea and weather conditions. (*Photo courtesy of National Data Buoy Center.*)

Here g is the gravitational constant, m the mass of the projectile (e.g., a stone,) and v_x and v_y are the x- and y-components of the projectile's velocity, respectively. The second derivatives d^2x/dt^2 and d^2y/dt^2 are the x- and y-components of acceleration. The launch speed V_0 and launch angle θ are set by the user. As suggested by Figure 2.2, the user

Figure 2.2. Catapult sends a projectile along a trajectory toward a target. Computing the parameters needed to hit the target is an example of analysis. Determining how to best build the catapult is an example of design.

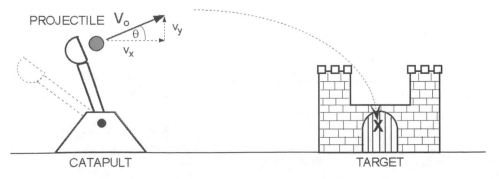

first must choose the target point, then adjust V_0 and θ so that the stone hits the desired target. Although this process involves making decisions (which target shall I hit?) and setting parameters (which V_0 and θ shall I choose?), and has more than one possible solution, it requires analysis only and involves no design.

In contrast, determining how to *build* a catapult capable of launching the stone at a desired velocity and angle clearly involves design. The machine can be built in more than one way, and the designer must decide which method is best. Should the structural members be made from oak or acacia wood? Should twisted ropes, a basket of rocks, or bent branches be used as the energy storage mechanism? How large should the machine be? Should it rest on wheels or skids? Answering these questions requires experimentation, analysis, testing, evaluation, and revision, which are all elements of the design process.

Reproduction

Design involves the creation of a device or product to meet a need or set of specifications. Analysis refers to the process of applying mathematics and other tools to find the answer to a problem. The word *reproduction* refers to the process of recreating something that's already been designed. Reproduction may involve exact replication or perhaps minor revisions whose consequences already have been determined. For instance, copying an oscillator circuit from an electronics book and substituting resistor values to set the frequency is an example of reproduction rather than design. Reproduction is an important part of engineering and lies at the core of manufacturing, but it does not require the same set of skills and tools as does true design.

Exercises 2.1

E1. Determine whether the following task involves analysis, design, or reproduction: Find the best travel route between your place of residence and the nearest airport.

E2. Determine whether the following task involves analysis, design, or reproduction: Find a way to prevent customers from burning their hands on paper cups of hot liquid purchased over the counter from a coffee shop.

E3. Determine whether the following task involves analysis, design, or reproduction: Find a way to stack cans of beans so that the smallest sized box can be used for shipping.

E4. Determine whether the following task involves analysis, design, or reproduction: Find a way to mount a cellular telephone inside an automobile to permit hands-free operation.

E5. Determine whether the following task involves analysis, design, or reproduction: Find a way to rapidly produce pom-poms that can be handed out to the fans at the Home U. versus Visitors football game.

E6. Determine whether the following task involves analysis, design, or reproduction: Find a way to rapidly produce origami (folded paper) nut containers for 400 people at a large alumni dinner.

E7. Determine whether the following task involves analysis, design, or reproduction: Find a way to use a global positioning system (GPS) receiver to automatically drive a vehicle between two preset locations.

2.3 GOOD DESIGN VERSUS BAD DESIGN

Anyone who has taken a car to an auto mechanic recognizes the difference between good and bad mechanics. A good mechanic diagnoses your problem in a timely manner, fixes what's broken at a fair price, and makes a repair that will last. A bad mechanic fails to find the real problem, masks the symptoms with expensive solutions that do not last, and charges too much money for unneeded repairs. Engineers are a bit like auto mechanics in this respect—some are good and some are bad. Just because an engineer produces something does not mean that the product has been well designed. Although the criteria by which a product is judged will vary with the nature of the product, most can be judged by the general characteristics summarized in Table 2.1.

TABLE 2-1 Characteristics of Good Design versus Bad Design

GOOD DESIGN	BAD DESIGN
1. Works all the time	1. Works initially, but stops working after a short time
2. Meets all technical requirements	2. Meets only some technical requirements
3. Meets cost requirements	3. Costs more than it should
4. Requires little or no maintenance	4. Requires frequent maintenance
5. Is safe	5. Poses a hazard to user
6. Creates no ethical dilemma	6. Fulfills a need that is questionable

The contrast between good and bad design is readily illustrated by the catapult example previously introduced. Consider the catapult shown in Figure 2.3. Suppose that Apex Catapult Co. has been asked to produce this device (actually called a trebuchet) for a brigade intent on rescuing their captured king and queen. The buyers will judge the worthiness of the catapult based on the considerations outlined in Table 2.1, as illustrated by the following discussion:

1. Does the Product Work?

A product under development need not work the first time it is tested, but it must work perfectly and repeatedly before it is delivered. It must be durable and not fail after only a short time in the field. Such a catapult provides an excellent example of this principle. Even a bad designer could produce a catapult capable of meeting its specifications upon initial delivery. The catapult might be made from whatever local timbers were harvested from the woods. It might utilize a simple trigger mechanism made from vines and twigs. The bad designer would build the catapult as he went along, adding new features on top of old ones without examining how each feature interacted with those before it. The catapult would likely pass inspection upon delivery and be able to hurl stones several times before fraying a line, cracking a timber, or jamming it's trigger mechanism. After a short period of use, the ill-designed timbers of its launch arm might weaken, causing the projectile to fall short of its mark.

A good designer would develop a robust catapult capable of many long hours of service. This conscientious engineer would test different building materials, carriage configurations, trigger mechanisms, and launch arms before choosing materials and design strategies. The catapult would be designed as a whole, with consideration given to how its various parts interacted. The process typically would involve many test cycles and the reworking of components to identify weak points. Building a battle-worthy catapult capable of heavy use in the field would require stronger and more expensive materials, but would prove more reliable in the long run and allow the user to hit the target repeatedly.

2. Does the Product Meet Technical Requirements?

It might seem a simple matter to decide whether a catapult meets its technical requirement. Either the stone hits its target, or it does not. But success can be judged in many

Figure 2.3.   Reproduction of a medieval trebuchet. (*Photo courtesy of Middelaltercentret.*)

ways. A catapult that represents good design will accommodate a wide range of stone weights, textures, and sizes. It will require the efforts of only one or two people to operate, and will repeatedly hit its target, even in strong wind or rain. A poorly designed catapult may meet its launch specifications under ideal conditions, but it may accommodate stones of only a single weight or require that only smooth, harder-to-find stones be used. When the arm of a poorly designed catapult is released, it may hit the carriage, causing the stone to lose momentum and fall short of the target.

3. Does the Product Meet Cost Requirements?

Some design problems can be solved without regard to cost, but most of the time cost is a major factor in making design decisions. Often a tradeoff exists between adding features and adding cost. A catapult made from cheap local wood will be much less expensive than one requiring stronger, imported wood. Will the consumer be willing to pay a higher price for a stronger catapult? Durable leather thongs will last longer than links made of less expensive hemp rope. Will the consumer absorb the cost of a more durable product? Painting the catapult will make it visually more attractive but will not enhance performance. Will the customer want an attractive piece of machinery at a higher price? A designer must face similar questions in just about every engineering endeavor.

4. Will the Product Require Extensive Maintenance?

A durable product will provide many years of flawless service. Durability is something that must be planned for as part of the design process. At each step, the designer must decide whether cutting corners to save money or time will lead to component failure later on. A good designer will eliminate as many latent weaknesses as possible. A bad designer will ignore them as long as the product can still pass its specification tests. If Apex Catapult Co. wishes to make a long-lasting product worthy of its company name, then it will design durability into its catapult from the ground up.

5. Is the Product Safe?

Safety is a quality measured only in relative terms. No product can be made completely hazard free, so when we say a product is "safe," we mean that is has a significantly smaller chance of causing injury than does a product that is "unsafe." Assigning a safety value to a product is one of the harder aspects of design, because adding safety usually requires adding cost. Also, accidents are subject to probability and chance, and it can be difficult to identify a hazard until an accident occurs. An unsafe product may never cause harm for a particular user, while a statistical fraction of many users may sustain injury. The catapult provides an example of the safety-vs.-cost tradeoff. Can a catapult be designed that provides a strategic advantage without injuring people? When a stone is thrown at a castle wall, a probability exists that it will hit a person instead. Designing a device that can throw, say, large bags of oil instead of stones would reduce this hazard to people, but at the added cost of hard-to-find oil. Features also could be added to the catapult to protect its users. Guards, safety shields, and interlocks would prevent accidental misfirings, but would add cost and inconvenience to the unit.

6. Does the Product Create an Ethical Dilemma?

The catapult has been chosen as an example for this section because it poses a common ethical dilemma faced by engineers: Should a device be built just because it *can* be? When asked to build a catapult, does Apex Catapult Co. have the responsibility to suggest alternatives to the rescue brigade? A less destructive battering ram might help save the king and queen while sparing innocent lives. Perhaps quiet diplomacy will lead to resolution and peaceful cooperation? As contrived as this fictitious example may be, it exemplifies the ethical dilemmas faced by engineers all the time. If asked by a future employer to design offensive military weapons, would you find it personally objectionable? If your boss asked you to use cheaper materials but bill the customer for more expensive ones, would you comply with these instructions or defy your employer? If you discovered a serious safety flaw in your own product that might lead to human injury, would you insist on costly revisions that would reduce the profitability of the product? Or would you say nothing and hope for the best? Questions of this type are never simple, but engineers must be prepared to answer them. As part of your training as an engineer, you must learn to apply your own ethical standards, whatever they may be, to problems that you encounter on the job. This aspect of design will be one of the hardest to learn but is part of your professional responsibility as an engineer.

Choose a Good Designer for a Mentor

In this section, we've highlighted the differences between good design and bad design. Practicing engineers of both types can be found in the engineering profession, and it's important that you learn to distinguish between the two. As you embark upon your transformation from student to professional engineer, you are likely to seek a mentor at some point in your career. Be certain that the individual you choose practices good design. Seek an engineer who has an intrinsic feeling for why and how things work. Find someone who adheres to ethical standards that are consistent with your own. Avoid "formula pluggers" who memorize equations and blindly plug in numbers to arrive at design decisions but have little feeling for what the formulas mean. Likewise shun engineers who take irresponsible shortcuts, ignore safety concerns, or choose design solutions without thorough testing. Do emulate engineers who are well respected, experienced, and are practiced at design.

2.4 THE DESIGN CYCLE

Design is an iterative process. Seldom does a finished product emerge without changes along the way. Sometimes, an entire design approach must be abandoned and the product redesigned from the ground up. The sequence of events leading from idea to finished product is called the *design cycle*. Although the specific steps of the design cycle may vary with the product and field of engineering, most cycles resemble the sequence depicted in Figure 2.4. The following sections explore this diagram in more detail.

Define the Overall Objectives

You should begin any new project by defining your design objectives. This step may seem a nuisance to the student eager to build and test, but it is an important one. Only by viewing the requirements from a broad perspective can an engineer determine all factors relevant to the design effort. Good design involves more than making technical choices. The engineer must ask the following questions: Who will use the product? What are the needs of the end user? What will the product look like? Which of its features are critical, and which are only desirable? Can the product be manufactured easily? How much will it cost? What are the safety factors? Who will decide how much risk is acceptable? Answering these questions at the outset will help at each subsequent stage of the design process.

Gather Information

At the early stages of a new project, you should devote time to gathering information. Learn as much as possible about technology related to the project. Identify useful off-the-shelf items that can be incorporated into your design. Send for catalogs, data sheets, and product information. Keep this information in a file folder where you'll find it easily. Also look for reports, application notes, or project descriptions in the same general area as your own project. A Web search can be extremely helpful in this respect, but filter information with care. Just because information has been posted on a Web site does not necessarily mean that it is accurate. Stick to official company Web sites or other reliable sources for technical information and application notes. View information from unknown sites with possible skepticism until you can verify its accuracy on your own.

Full product information, often more comprehensive than that available on the Web, usually can be obtained directly from the manufacturer. Perusing advertisements in trade magazines and journals can be a good way to learn what types of products are available in your field of interest. Often the magazine itself has a master reply card on which you can circle several requests for product information. Each field of engineering has many such publications. A few examples include *Electronic Design News* (electrical); *Compliance Engineering* (electrical); *Mechanical Engineering* (mechanical); and *Machine Design* (mechanical).

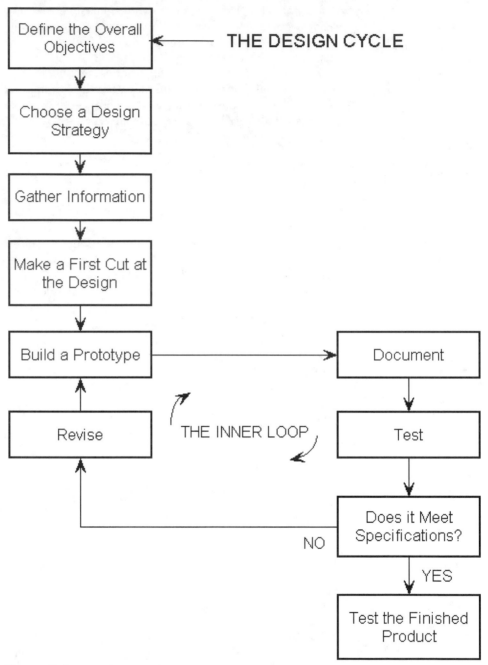

Figure 2.4. The design cycle.

Choose a Design Strategy

The next step in the design process involves the selection of a design strategy. This strategy will define the *methods* that you'll use to meet your design objectives. At this stage, the engineer (or, more likely, the design team) might decide whether the design will involve an electrical or mechanical solution, or whether the product will be synthesized from off-the-shelf components or from basic raw materials. If the system is complex, it should be broken up into simpler, smaller pieces that can be designed independently and later interconnected to form the complete product. These subsections, or *modules*, should be designed so that they can be tested individually before the entire system is assembled. Subdividing a large job into several more manageable tasks simplifies synthesis, testing, and evaluation. In a team design effort, the modular approach is essential. The various components of an automobile, for example, such as the engine, cooling system, electrical system, suspension, braking system, chassis, and drive train, are each tested individually before the entire automobile is assembled.

The design strategy also should consider any similar designs that are ongoing or have been created in the past. Perhaps a new technology exists that will be useful for your design? It may be that a partial solution is already available in commercial form. The wise engineer makes use of existing products and components to simplify the design task. There is no shame in using off-the-shelf ingredients or subsystems if they can help you achieve your design objectives more quickly and inexpensively. Typically, labor is the most expensive part of any development effort, so it's often cheaper (and sometimes more reliable) to buy something ready-made than to design it from scratch. For example, imagine how needlessly complex the task of designing a desktop computer would be without making use of the disk drives, memory chips, power supplies, monitors, and central processors available from other vendors. One caution: Be certain that using another company's product does not create patent infringement problems if yours is destined for commercial sale.

Make a First Cut at the Design

After the design strategy has been solidified, it is time to make an a first cut at the design. If the product is to be a physical entity, a detailed layout or engineering drawing is helpful at this stage. Tentative values of dimensions and other parameters should be assigned wherever possible. Key sizes, dimensions, construction materials, part numbers, electronic component values, and other relevant parameters are specified. This step typically involves *rough approximations* and gross estimates. Its primary purpose is to determine whether or not the design approach has a chance of working, and it should result in a rough, tentative prototype for the system or each of its subsections. If the object of the design effort is a software product, its outer shell, overall structure, and user interface are laid out as part of the first-cut attempt.

Build, Document, and Test

After the first cut has been committed to paper, it's next translated into a working prototype. The first prototype will be revised many times before the design cycle has been completed. It should be functional, but need not be visually attractive. It's primary purpose is to provide a starting point for evaluation and testing. If the product is software, for example, the prototype might consist of the main calculation sections without the fancy graphical interfaces that the user will expect in the final version. If the product is mechanical, it can be built up as a mock-up from easy-to-modify materials, such as wood or prepunched metal strapping. For instance, a prototype for a new washing machine engine might be built inside an open wooden box made from framing lumber and hinged plywood. Such a structure would permit easy access to the inner machinery during testing but obviously would never make it to the sales floor. If the product is electrical, its prototype could be wired on a temporary circuit breadboard. A new digital clock, for example, might be built in this way so that the designer could have easy access to its timing and display signals for test purposes. This arrangement would allow for circuit revisions before final construction and packaging.

During the prototyping phase of design, computer simulation tools such as AutoCAD™, ProENGINEER™, SPICE, or Simulink™ can save time and expense by allowing you to predict performance before actual construction of the prototype. These software packages can help you identify hidden flaws *before* the product is built and give you some indication as to the success of your design approach. However, they should never be used as a substitute for physical testing. Glitches, bugs, and other anomalies caused by physical effects not modeled by the simulator have a nasty habit of appearing when the real product is tested. Despite the usefulness of computer-aided design tools and simulators, there is simply no substitute for constructing and testing a real physical prototype.

Note that documentation is part of the inner loop of the design cycle of Figure 2.4. The typical engineer faces considerable temptation to leave documentation to the very end. Pressed with deadlines and project milestones, many an inexperienced engineer thinks of documentation as a nuisance and an intrusion rather than as an integral part of the design process. After working diligently on a design project, this engineer then faces the dread realization, "Oh, my gosh, now I've got to write this up!" Documentation added as an afterthought is often incomplete or substandard, because most of the relevant facts and steps have been forgotten by the time writing takes place. Haphazard, after-the-fact documentation is the cause of many a design failure. Many a product, developed at great cost, but delivered with pathetic documentation, has found its way to the trash heap of engineering failures because no one could figure out how to use or repair it. Poor documentation also leads to duplication of effort, or reinvention of the wheel, because no one can remember or interpret the results of previous work.

A good engineer recognizes that documentation is absolutely critical to every step of the design process. He or she will plan for it from the very beginning, keeping careful records of everything from initial feasibility studies to final manufacturing specifications. It's a good idea to write everything down, even if it seems unimportant at the time. Documentation should be written in such a way that another engineer who is only slightly familiar with the project can pick up the work at any time by simply reading the documentation. Also, careful documentation will aid in writing product literature and technical manuals should the product be destined for commercial sale. Good documentation provides the engineer with a running record of the design history and the answers to key questions that were asked along the way. It provides vital background information for patent applications, product revisions, and redesign efforts and serves as insurance in cases of product liability. Above all, documentation is part of an engineer's professional responsibility, and it's importance to engineering design cannot be overemphasized. Because the issue of documentation is so important, we shall revisit it in more detail in later chapters.

Revise and Revise Again

One of the characteristics of design that distinguishes it from reproduction is that the finished product may be totally different from what was envisioned at the beginning of the design cycle. Elements of the system may fail during testing, forcing the engineer to rethink the design strategy. The design process may lead the engineer down an unexpected path or into new territory. A good engineer will review the status of a product many times, proceeding through numerous revisions until the product meets its specifications. In truth, this revision process constitutes the principal work of the engineer. An experienced engineer recognizes it as a normal part of the design process and does not become discouraged when something fails on the first or second try. The revision cycle may require many iterations before success is achieved.

Thoroughly Test the Finished Product

As the design process converges on a probable solution, the product should be thoroughly tested and debugged. Performance should be assessed from many points of view, and the design should be modified if problems are identified at any stage. If the product is a physical entity, the effects of temperature, humidity, loading, and other environmental factors, as well as the effects of repeated and prolonged use all must be taken into account. A physical product should be subjected to an extended "burn-in," or use period, to help identify latent defects that might cause it to fail in the field. The human response to the product also should be assessed. No two people are exactly alike, and exposing the product to different individuals will help identify problems that may not have been apparent during the development phase. Similarly, a software product should be tested at numerous "beta sites" so that a variety of different users can weed out hidden bugs. Only after a comprehensive test period is the product ready to be put into actual service.

PROFESSIONAL SUCCESS: HOW TO TELL A GOOD DESIGN ENGINEER FROM A BAD DESIGN ENGINEER

As you pursue your engineering career, you will encounter many individuals. Some will be good engineers, and others will be bad engineers. In your quest to identify and emulate only good engineers, you should learn the differences between the two. The following list highlights the traits of both types of engineers:

A Good Engineer:

- Listens to new ideas with an open mind.
- Considers a variety of solution methodologies before choosing a design approach.
- Does not consider a project complete at the first sign of success, but insists on testing and retesting.
- Is never content to arrive at a set of design parameters solely by trial and error.

- Uses phrases such as "I need to understand why," and "Let's consider all the possibilities."

A Bad Engineer:

- Thinks he/she has all the answers; seldom listens to the ideas of others.
- Has tunnel vision; pursues with intensity the first approach that comes to mind.
- Ships the product out the door without thorough testing.
- Uses phrases such as, "good enough" and "I don't understand why it won't work; so-and-so did it this way."
- Equates pure trial and error with engineering design.

Exercises 2.2

E8. Without looking at Figure 2.4, draw a diagram of the design cycle that includes the following steps: *define, gather, choose, first cut, build, document, test, revise.*

E9. Make a list of the various ways in which an engineer might gather information as part of the design cycle.

E10. Draw a modified design cycle, similar to Figure 2.4, that includes feedback from a test group of individual users.

E11. Define the design strategy that might have gone into the invention of the wheel.

E12. Describe the various elements of the design cycle as they might have applied to the development of the common paper clip.

2.5 A DESIGN EXAMPLE

The principles presented in Section 2.4 describe the essence of the design process. Let's illustrate these steps using a simple example. In this section, we'll introduce a specific design problem and use it to follow the phases of the design cycle. We will revisit the same design problem in later chapters as we explore elements of the design process in more detail.

The Peak Performance Design Competition

Suppose that your professor has announced a design competition as part of an introductory engineering course. The rules of the contest are outlined in the following flyer:

COLLEGE OF ENGINEERING
PEAK-PERFORMANCE DESIGN COMPETITION

OBJECTIVE

The goal of the competition is to design and construct a vehicle that can climb a ramp under its own power, stop at the top of the ramp, and sustain its position against an opposing vehicle coming up the other side of the ramp. The illustration of Figure 2.5 shows the approximate dimensions of the ramp. The 30-cm width of the carpet-covered track may vary by ±0.5 cm as the vehicle travels from the bottom to the top of the ramp. A vehicle is considered to be on "top of the hill" if it, including any extensions, strings, or jettisoned objects, is completely within the two 120-cm lines at the end of a 15-second time interval. (See illustration.)

VEHICLE SPECIFICATIONS

1. The vehicle must be autonomous. No remote power, control wires, or remote control links are allowed.

2. The vehicle's exterior dimensions at the start of each run must not exceed 30 cm × 30 cm × 30 cm. A device, such as a ram, may extend beyond this limit once activated, but cannot be activated before the start of the run.

3. The vehicle must be started by an activation device (e.g., switch, mechanical release, etc.) on the vehicle. Team members may not activate any device before the start. Vehicles cannot be running and dropped to start.

4. The vehicle can be powered only by one of the following energy sources:
   - Batteries of up to 9 volts
   - Rubber bands (4 mm × 10 cm maximum size in their unstretched state)
   - Mouse traps (spring size 1 cm × 3 cm maximum)

5. The vehicle's weight, including batteries, must not exceed 2 kg.

6. The vehicle must not use chemicals or dangerous substances. No rocket-type devices, CO_2 propulsion devices, or chemical reactions are allowed. No mercury switches are permitted. Mercury is a toxic substance, and a risk exists that a mercury switch will break during the competition.

7. The vehicle must not be anchored to the ramp in any way before the start. At the end of the run, the vehicle and all its parts, including jettisoned objects, extensions, etc., must lie completely within the top of the hill and the 30-cm track width.

8. The vehicle must run within the 30-cm wide, carpet-covered track. It may not run on top of the guide rails.

9. The vehicle must compete in six 15-second runs against opposing vehicles. The vehicles with the most wins after six runs will be selected for the Grand Finale, which will determine the winner of the competition. Modifications to the vehicle are permitted between (but not during) runs.

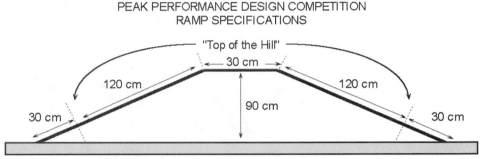

Figure 2.5. Ramp specifications for the Peak Performance Design Competion.

Let's now examine this problem statement in the context of the design cycle presented in Figure 2.4. Remember that the problem can be addressed in any number of ways, but some design solutions will work better than others.

Define the Overall Objectives

When faced with the task of designing something, many an engineer is tempted to begin with construction right away, before a careful planning stage. Building things is fun and satisfying, while estimating, sketching, calculating, simulating, and checking design parameters seems less glamorous. Its important, however, to begin any project by taking time to define its objectives.

Suppose that you and a classmate have decided to enter the design competition. Your first step should be to write down a list of general criteria to be met by your design effort. You make a list of tradeoffs that characterize the problem as you try to foresee problematic areas and develop strategies for overcoming them. You arrive at the following set of design objectives:

1. *Design for speed.* The fastest vehicle will not necessarily be the winner, but a in order to win, the vehicle must reach and maintain the center line well before the 15-second time limit.

2. *Design for defensive and offensive strategies.* Not only must your vehicle reach the top of the ramp and stop on its own, but it must maintain its position as your opponent tries to do the same. Although offensive and defensive strategies are not necessarily mutually exclusive, you've decided (somewhat arbitrarily) that defense will be given a higher priority than offense.

3. *Design for easy changes.* The rules state that modifications to the vehicle are permitted between runs. During the contest, you may see things on other vehicles that will prompt you to make changes in your own vehicle. Adopting an easy-to-change construction strategy will facilitate on-the-fly changes to your vehicle. The likely tradeoff in choosing this approach is that your car will be less durable and more likely to suffer a disabling failure.

4. *Design for durability.* The vehicle must endure six, and possibly more, trips up the contest ramp. Opposing vehicles and accidents can damage a fragile design. You must weigh the issue of durability against your desire to produce a vehicle that's flexible and easy to modify.

5. *Design for simplicity.* By keeping your design simple, you will be able to repair your vehicle quickly and easily. An intricate design might provide more performance features, but it also will be more prone to breakdowns and will be more difficult to repair.

Note that these goals are not independent of one another. For example, designing for easy changes may conflict with building a durable vehicle. Designing for both offensive and defensive strategies will lead to a more complicated vehicle that is harder to repair. Engineers typically face such tradeoffs when making design decisions. Deciding which pathway to take requires experience and practice, but making any decision at all means that you've begun the design process.

Choose a Design Strategy

Many different design strategies will lead to a vehicle capable of competing. Building a *winning* design, however, requires careful planning and attention to numerous details and design features. Success requires making the right choices at each step in the design process. How can you know ahead of time what the right choices will be? In truth, you cannot know, especially if you have never built such a vehicle before. You can only make educated guesses based on your experience and intuition. You test and retest your design choices, making changes along the way if they increase your vehicle's performance. This process of *iteration* is a crucial part of the design process. Iteration refers to the process of testing, making a change, and then retesting to observe the results. Good engineering requires many iterations, trials, and demonstrations of performance before a design effort is completed. In the world of engineering, the first cut at a design seldom resembles the finished product.

The rules of the competition provide for many alternatives in vehicle design. Regardless of the details of the design, however, all vehicles must have the same basic components: *power source, propulsion mechanism, stopping mechanism,* and *starting device.* Although not required, a defense mechanism that prevents an opponent from pushing the vehicle back down (or off) the ramp will increase your chances of winning. After some discussion with your teammate, you develop the *choice map* shown in Figure 2.6. This diagram displays some of the many design choices available to you. Although it does not provide a complete list, it serves as an excellent starting point for your design attempt. As your teammate points out, "We have to start somewhere." You remark in return, "Let's begin with our best guesses about which of these choices will work." The choice map of Figure 2.6 includes the following elements:

Power Source   According to the rules of the competition, you may power your vehicle from standard nine-volt (9V) batteries, rubber bands, or mousetrap springs. Batteries are attractive because they require no winding or preparation other than periodic replacement. They will, however, be more expensive than the two alternatives. Rubber bands will require much less frequent replacement, but will store the least amount of energy among the three alternatives. Like a rubber band, a mousetrap needs no frequent replacing. It stores more energy than a rubber band, but offers the fewest options for harnessing its stored energy.

CHOICE MAP FOR PEAK PERFORMANCE VEHICLE

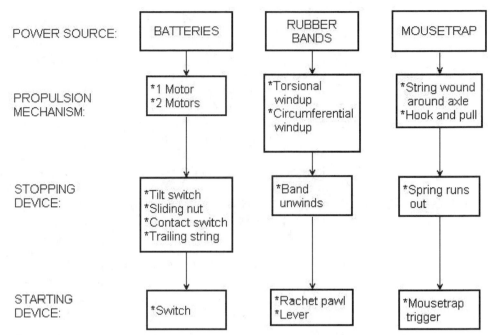

POWER SOURCE:	BATTERIES	RUBBER BANDS	MOUSETRAP
PROPULSION MECHANISM:	*1 Motor *2 Motors	*Torsional windup *Circumferential windup	*String wound around axle *Hook and pull
STOPPING DEVICE:	*Tilt switch *Sliding nut *Contact switch *Trailing string	*Band unwinds	*Spring runs out
STARTING DEVICE:	*Switch	*Rachet pawl *Lever	*Mousetrap trigger

Figure 2.6. Choice map that outlines the decision tree for the first phase of the design process.

Propulsion Mechanism   Your choice of the propulsion device, or *prime mover,* will depend entirely on your choice of power source. If a battery is used as the power source, an electric motor seems the obvious choice for turning the vehicle's wheels. Rubber bands can be stretched to provide linear motion or twisted for torsional energy storage and used to turn an axle or power shaft. Alternatively, a rubber band can be stretched around a shaft or spool like a fishing reel and used to propel the vehicle's wheels. A mousetrap can provide only one kind of motion. When released, its bale will retract in an arc, as depicted in Figure 2.7. This motion can be harnessed and used to propel the vehicle.

Stopping Mechanism   Your vehicle must stop when it arrives at the top of the ramp. This requirement can be met by interrupting propulsion power precisely at the right moment. A braking device to augment power shutoff should also be considered. If the

Figure 2.7. Harnessing the stored mechanical energy of a mousetrap. The bale retracts in an arc when the mousetrap is released.

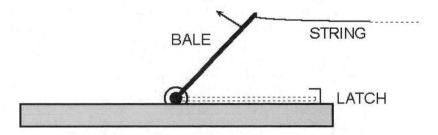

vehicle is powered by batteries, there are many possibilities for a stopping device. A simple tilt switch that disconnects the battery when the vehicle is level, but connects the battery when the vehicle is on a slope, will certainly do the job. For safety reasons, however, the rules prohibit the use of mercury switches (elemental mercury is toxic to humans), so any tilt switch used in the vehicle must be of your own design. One interesting concept would be to use an electronic timer circuit that shuts off power from the battery after a precise time interval. Through trial and error, you could set the elapsed time to just the right value so that the vehicle stops at the top of the ramp. One problem inherent to this open-loop timing system is that the vehicle does not actually sense its own arrival at the top of the hill, but rather infers it by precise timing. The possibility exists that the speed of the vehicle may be inconsistent from run to run, especially if the vehicle is battery powered. As battery energy is depleted, the speed of the vehicle will decrease, requiring more time to reach the top of the hill. Another alternative might be to connect a mechanism that cuts off power to the wheels after the car has traveled a preset distance as measured by wheel rotations. This scheme will work well if the wheels do not slip.

If a rubber band or mousetrap propulsion system is chosen, then stopping the vehicle will require something other than an electrical switch. One crude way of stopping a vehicle propelled by mechanical energy storage is simply to allow the primary power source to run out (e.g., by allowing the rubber band to completely unwind). However crude this method, it is reliable, because power input to the vehicle will *always* cease when the source of stored energy has been depleted.

Starting Device   If the vehicle is powered by a battery, then an electrical switch becomes the most feasible starting device. A rubber band power source will require a ratchet pawl, trip lever, or other similar device in order to initiate power flow to the wheels. A mousetrap can make use of its built-in trigger mechanism or any other starting mechanism that you can devise.

Make a First Cut at the Design

The first design iteration begins with rough estimations of the dimensions, parameters, and components of the vehicle to make sure that the design is technically feasible. After discussing the long list of design choices, you and your teammate decide upon a battery-powered vehicle. This decision makes available many choices for the stopping device and power train that far outweigh the advantages of mechanical propulsion schemes. You decide upon a defensive strategy and agree to build a slower-moving, wedge-shaped vehicle driven by a small electric motor. The advantage of this design strategy is that the motor can be connected to the wheels using a small gear ratio, thereby providing higher torque at the wheels and a mechanical advantage unavailable to a very fast vehicle. You plan to use plastic gears and axles purchased from a hobby shop. The gear box will reduce the speed of the wheels relative to the motor shaft speed, providing added mechanical torque that will significantly increase the force available to push the opposing vehicle off the ramp. Because your vehicle will be slower than others, it may not reach the top of the ramp first, but its wedge shaped design will help to dislodge the front of any opposing vehicle that arrives first at the top of the hill. If your vehicle should happen to arrive first at the top of the hill, your car's defensive wedge shape will cause your opponent's car to ride over your car's body, allowing you to maintain your place in the center of the ramp.

You decide to use one motor with a single driven axle attached to both rear wheels. An alternative strategy would be to drive each rear wheel separately, thereby allowing the driven wheels to turn at different speeds. Such *differential* capability is essential for vehicles that travel curved paths, but in this case the vehicle must travel along a straight path only. By driving the wheels from a common shaft, you will reduce slippage, because *both* wheels will have to lose traction before forward motion will be impaired. You briefly consider front-wheel drive, because you assume from listening to many car advertisements that front-wheel drive is superior to rear-wheel drive. Your teammate is quick to point out that the advantage of front-wheel drive for automobiles lies in its ability to help the car negotiate curves. Despite the media-driven message of "better traction," the advantages of front-wheel drive have nothing to do with your application. In fact, front-wheel drive may be a disadvantage to your wedge-shaped design, because it may cause your car to flip over forwards if another car travels on top of it. Your teammate draws the sketch of Figure 2.8 to illustrate this scenario. You abandon the idea of front-wheel drive.

Build, Document, Test, and Revise

A rough preliminary sketch of your car is shown in Figure 2.9. You've entered this sketch into a notebook in which you've been recording all information relevant to the project. Included in your notebook are design calculations, parts lists, and sketches of various pieces of the car. Shown in Figure 2.9 are the car's wedge-shaped design, a single drive shaft driven by a motor, belt, and pulleys, and a single switch to turn off the motor when the vehicle arrives at the top of the hill. Your design concept represents a tradeoff between several competing possibilities, but you and your teammate have decided that the car's electric motor drive and defensive shape have the best chance of winning the competition.

Figure 2.8.   Front and rear wheel drive options for a moving wedge vehicle.

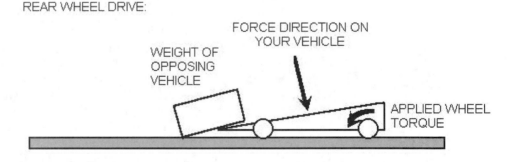

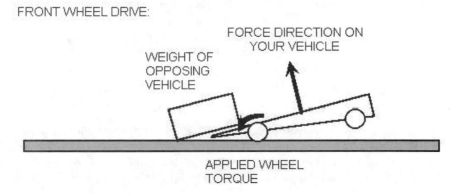

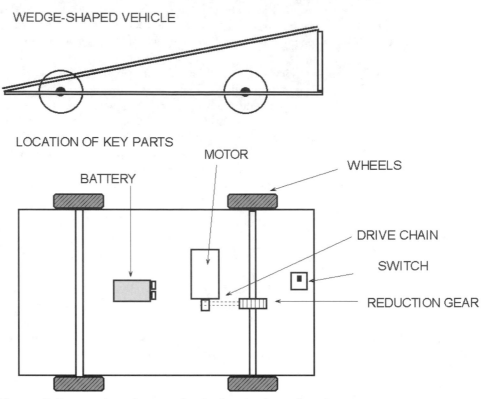

Figure 2.9.   Rough, preliminary sketch of car for the Peak Performance Design Competition.

The sketch in Figure 2.9 represents a beginning, but it is not the finished product. You still have many hurdles to overcome before the design is ready to compete. The next step in the design process involves building and testing a first-cut prototype. To help you in this phase of the design process, your professor has constructed a test ramp available to all contestants. You begin by constructing a chassis shell in the form of a wedge, but without a motor drive or stopping mechanism.

You run your wedge-shaped vehicle up the ramp by hand. You soon discover that the bottom of the vehicle hits the ramp at the top of the hill, as depicted in Figure 2.10(a). The change in angle of the ramp is large, and all four wheels do not always maintain contact with the track surface. You discuss several solutions to this problem with your teammate. One solution would be to increase the size of the wheels, as shown in Figure 2.10(b). This change would decrease the mechanical advantage between the motor and the wheels, requiring you to redo your calculation of how much force will be required from the motor. Another solution would be to make the vehicle shorter, as shown in Figure 2.10(c), but you realize that this solution would lead to a steeper angle for the wedge shape of the vehicle and reduce its effectiveness as a defensive strategy. (The largest thickness of the vehicle must stay the same to accommodate the motor and gear box.)

Revise Again

Your teammate suggests keeping the wheels and shape of the wedge the same and simply moving the rear wheels forward, as depicted in Figure 2.11. You rebuild the vehicle by moving the rear shaft mount forward, and you test the vehicle again. The redesigned vehicle no longer bottoms out on the track, and you claim success. Your professor sees

VEHICLE HITS RAMP

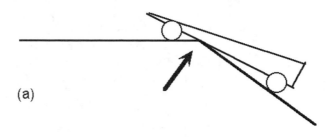

(a)

LARGER WHEELS

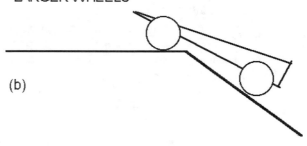

(b)

MAKE VEHICLE SHORTER

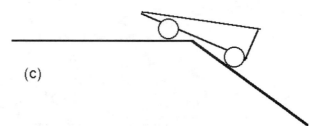

(c)

Figure 2.10. Vehicle at the top of the ramp. a) Bottom of vehicle hits the ramp; b) vehicle with larger wheels; c) a shorter vehicle.

REAR WHEELS
MOVED FORWARD

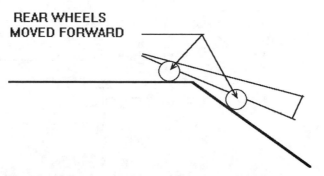

Figure 2.11. Moving the rear wheels forward.

your design changes and suggests that you test your vehicle under more realistic conditions. For example, what will happen when another vehicle rides over the top of your wedge-shaped body? You proceed to simulate such an event by placing a weight at various positions on the top of the car. The results of these additional tests suggest that the wheel location modification may not be the best solution to your problem. When you move the rear wheels forward, you change the base of support for the car's center of gravity. You discover that if an opposing vehicle rides over the top of your car, the net center of gravity moves toward the rear, eventually causing your car to topple backwards, as depicted in Figure 2.12.

These discoveries and setbacks may seem discouraging, but they are a normal part of the design process. Some things work the first time, while others do not. By observing and learning from failure and by building, testing, revising, and retesting, you can converge on the best solution that will meet your needs.

After some thought, you decide that increasing the size of the wheels may be the best option after all. Your teammate points out that you can simply change the ratio of the gear box to preserve the net mechanical advantage between the motor and the drive shaft. This change will allow you to accommodate larger wheels. You buy some new wheels and try them with success. With the rear axle moved to its original location and the larger wheels in place, your car no longer bottoms out on the track.

You next consider the motor that will power the drive shaft. Motors of all sizes and voltage ratings are available, including some alternating current (ac) motors, as well as direct current (dc) motors. Given that your car will be powered by batteries, your obvious choice is a dc motor. What voltage rating should you choose? You find no motors rated at 9 volts at the local hobby shop. "I don't think anyone makes 9-volt motors," says

Figure 2.12. Weight of opposing vehicle on top of rear end causes car to topple over backwards.

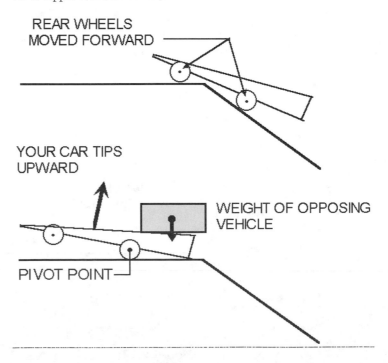

the salesperson behind the counter. You look up the Web pages of vendors*, scrutinize several catalogs , and find motors rated for 3, 6, 12, or 24 volts, but no 9-volt motors. Your professor explains that the rating of a motor specifies its operating voltage under steady-state use. If a lower-than-rated voltage is used, the maximum torque available from the motor will be reduced. If a higher-than-rated voltage is connected, the excess current will heat the windings inside the motor and possibly damage it. In the Peak Performance design competition, however, the motor will be energized only for about 15 seconds at a time. This interval may be short enough to allow a larger-than-rated voltage to be applied without harming the motor. The feasibility of this intentional overloading must be verified by testing or by contacting the motor's manufacturer.

After hearing your professor's explanation, your team decides to purchase several different motors rated at 3 volts and 6 volts. You test each one by connecting it to a variable voltage supply and measure the current flow at various applied voltages. You compute the power flow by multiplying voltage and current. You devise the apparatus shown in Figure 2.13 to measure the mechanical power delivered by the motor. Your contraption is a crude version of the industry-standard Prony brake used to measure motor torque. The frictional rubbing of the weighted loop of string applies a mechanical load to the motor. You power each of your motors at the same voltage and add weights until the motor stalls. The motor that sustains the largest weight before stalling will have the largest torque. You check your mechanical loading measurements against your electrical measurements and use your data to determine which motor gives you the most mechanical torque when powered by 9 volts.

As suggested by your professor, you and your teammate each keep a design notebook in which you record all your design decisions and the results of all your experiments and tests. Figure 2.14, for example, shows a page from your notebook in

Figure 2.13. Simple apparatus to measure motor torque.

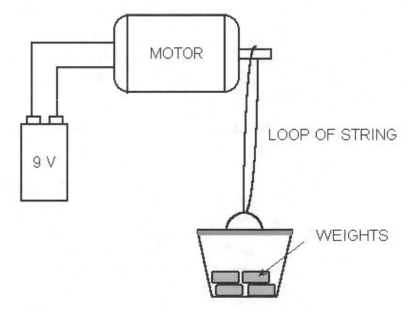

* See, for example, http://www.robotics.com; http://www.nwsl.com/products; http://www.morat.com; or http://www.sdp-si.com.

MOTOR A: TC-254		
# weights	V(volts)	I (amps)
0	9.0	0.02
1	8.9	0.05
2	8.9	0.07
3	8.8	0.1
4	8.7	0.2
5	8.0	0.5

MOTOR B: RS-257-234		
# weights	V(volts)	I (amps)
0	9.0	0.05
1	8.8	0.1
2	8.7	0.17
3	8.5	0.23
4	8.0	0.3
5	7.6	0.7

MOTOR C: BOD-37A		
# weights	V(volts)	I (amps)
0	9.0	0.01
1	8.9	0.04
2	8.9	0.05
3	8.8	0.1
4	8.7	0.25
5	8.0	0.52

MOTOR D: TOSH-7954		
# weights	V(volts)	I (amps)
0	9.0	0.01
1	8.9	0.11
2	8.9	0.21
3	8.8	0.31
4	8.7	0.41
5	8.7	0.51

Figure 2.14. Page from lab notebook that documents mechanical loading tests on various motors.

which you've recorded a list of your motor collection plus the results of the mechanical loading tests. You've also sketched the loading apparatus of Figure 2.13 in your notebook. It's important to record the characteristics of the motors you *do not* use in case your design needs change and you need to reconsider one of the rejected motors. As your design proceeds, you record all calculations, specifications, and sketches pertinent to the drive train, including gear ratios, electrical power consumption, wheel diameters, weight of each part, and construction techniques. Your objective is to have a complete record of your design activities by the time you enter the vehicle competition.

Exercises 2.3

E13. Make a two-column list that outlines the advantages of the various power sources for the Peak Performance Design Competition.

E14. Make a list of additional propulsion mechanisms not mentioned in this section that could be used to drive a Peak Performance vehicle.

E15. Make a two-column list that outlines the advantages of using gravity versus friction as a stopping mechanism for the Peak Performance Design Competition.

E16. Determine the minimum energy needed to lift a Peak Performance vehicle of maximum allowed weight from the bottom to the top of the ramp.

E17. How much electrical energy (in joules) is needed to exert 1 N of force over a distance of 1 meter?

E18. How much electrical power (in watts) is needed to exert 1 N of force on a body over a distance of 1 meter for 10 seconds?

E19. Determine the number of turns per cm of wheel diameter that will be required to move a vehicle from the bottom to the top of the ramp in the Peak Performance Design Competition.

E20. Determine the diameter of the wheels needed to move a vehicle from the bottom to the top of the ramp in the Peak Performance Design Competition with 50 turns of the drive axle.

SUMMARY

This chapter has outlined the essential elements of design at a very basic level. Design differs from analysis and reproduction because it involves multiple paths to solution, decision making, evaluation, and testing. The *design cycle* is an important part of engineering problem solving, as are knowledge, experience, and intuition. Documentation is critical to the success of a product and should be an integral part of the design process.

The chapters that follow examine several aspects of design in more detail using the Peak Performance Design Competition as a continuing example. Some of the topics to be presented include teamwork, brainstorming, documentation, estimation, modeling, prototyping, project organization, and the use of computers in engineering design.

KEY TERMS

Design	Analysis	Design cycle

Problems

The following problem statements can be used to practice problem solving and idea generation. Some of them involve paper designs, while others are suitable for actual fabrication and testing.

1. Develop a design concept for a mechanism to allow hands-free operation of a telephone. Outline its basic form, key features, proposed method of construction, and prototyping plan. Consider size, weight, shape, safety factors, and ease of use.

2. Design a device for securing a coffee cup near the driver's seat of an automobile. The device should prevent the cup from spilling and should not interfere with the proper operation of the car. It should be universally adaptable to a wide variety of vehicles. Address safety and liability issues as part of your design.

3. Design a device for carrying a cellular telephone on a bicycle. Your contraption should allow the rider to converse while holding on to the handlebars with both hands.

4. Develop at least three design concepts for a nonlethal mousetrap. Is your device cost effective compared with an ordinary, spring bale mousetrap? Is your trap more humane, and is it worth any added cost?

5. Devise a concept for a device that can dig holes in the ground for the installation of fence posts. Develop a prototype and test plan for your design concept.

6. Design a device that will allow the inside and outside surfaces of windows to be cleaned from the inside only. Compare projected cost with that of a simple, handheld, squeegee-type window cleaner.

7. Design a system for feeding aquarium fish automatically when the owner is out of town.

8. Develop a design concept for a spill-free coffee cup.

9. Design a device for carrying bricks to the top of a house for chimney repair. The alternative is to carry them up a ladder by hand.

10. Design at least three different systems for measuring the height of a tall building. If resources permit, try out your designs on a nearby building and compare results.

11. Design a system that will enable a self-propelled lawn mower to cut the grass in a yard unattended.

12. Design a system for minimizing the number of red lights encountered by cars traveling east and west through a major city. The system should not unnecessarily impede north-south traffic flow.

13. Devise a method for managing the flow of two-way railroad traffic on a one-track system. The single line of track has a parallel spur track located every five kilometers. These parallel sections of track allow one train to wait as another passes by in the opposite direction.

14. Design a public transportation system in which every traveler can ride a private vehicle on demand from any one station to another.

15. Develop a design concept for a system to measure the speed of a passing train.

16. Design a system for automatically turning off a small electric baking oven when a cake inside is done.

17. Design an electric switch that will turn off lights if the room is vacated.

18. Design a device that will enable a quadriplegic to change the channels on a television set.

19. Devise a system for turning on security lights at dusk and turning them off at dawn. These lights are to be installed throughout a large factory. You system should have a single master override for all lights.

20. Design a warning system for your home freezer that will alert you to abrupt changes in temperature.

21. Design a system for keeping a satellite's solar panels pointed at the sun.

22. Design a system that will automatically water houseplants when they are in need of moisture.

23. Design a device that will allow a one-armed individual to properly use dental floss.

24. Design a method for counting the number of people who attend a football game.

25. Design a lamp post that will break away when struck by an errant automobile.

26. Design an irrigation system that will bring water from a nearby pond to your vegetable garden.

27. Design a system that can aerate the pond in a city park so that algae growth will not overtake other forms of wildlife.

28. Design a system for automatically steering a ship along a desired compass heading.

29. Design an apparatus that will keep a telescope pointed at a distant star despite the rotation of the earth.

30. Devise a system for painting car bodies automatically by robot. You must include a method for training the robot for each painting task.

31. Devise a plan for a campus-wide information system that will allow any professor to access the grades of any student, while maintaining the privacy of the system to other users. A student also should be able to obtain his or her own grades, but not those of others.

32. Design an electric pencil sharpener that will turn off when the pencil has been properly sharpened.

33. Develop a design concept for a device that will turn on a car's windshield wipers automatically when rain falls. The wipers should come on only momentarily when rainfall is light but should be on all the time when the rainfall is moderate to heavy.

34. One of the problems with recycling of post-consumer waste is the sorting of materials. Consumers and homeowners cannot always be relied upon to sort correctly, yet one erroneously placed container can ruin a batch of recycled material. At the present time, most municipalities resort to manual labor to sort recyclable materials. Devise a concept that will sort metal cans, plastic bottles, and plastic containers at a recycling plant. Develop a plan for modeling and testing your system.

35. Design a concept for a voice-synthesized prompting system that can provide cues for an individual who must take medication on a strict regimen. The device need not be pocket sized, but should be portable and battery operated.

36. Design a system consisting of several panic buttons that will be installed at each of several workshop fabrication stations on a factory floor. Pressing any one of these buttons would activate a signal at a central control console and identify the location of the activated button. Ideally, voice communication over the system would be a desirable feature. One matter to consider is whether a wired or a wireless system is better.

37. An elementary school teacher needs a calendar-teaching system to help students learn about dates, appointments, and scheduling events. Your professor has asked you to develop a design concept. The basic system should be a large pad over which a monthly calendar can be placed. The underlying pad should have touch-sensitive sensors that can detect a finger placed on each day block in the calendar. The entire unit should interface with a desktop computer which will run a question-and-answer game or program. The typical types of questions to be asked might include, "You have scheduled a dentist appointment two weeks from today. Point to the day on the calendar on which you should go to the dentist," or "Sara's birthday is on February 11. Point to that day on the calendar." An appropriate reward, either visual, auditory, or both, should be issued by the computer for correct answers. A nonintimidating signal should be issued for incorrect answers. Outline the key features of your system and devise a development plan.

    The following three projects involve devices that will assist the professor running the Peak Performance Design Competition introduced in Section 2.5:

38. The rules of the Peak-Performance Design Competition state that each vehicle must be placed behind a starting line located 30 cm up the side of a 1.5-m ramp. After the starting signal is given from the judge, contestants must release their vehicles, which then have 15 seconds to acquire and maintain a dominant position on top of the ramp. Any vehicle that travels over the 30-cm starting line prior to the "go" signal loses the run. Currently, the starting sequence is initiated orally by the judge and timed by stopwatch. This system leads to much variability between judges, as many use different starting signals (e.g., "on your mark, get set, go" or "one, two, three, go"), and one may be lax in timing or checking for starting line violations.

    Design a system consisting of starting-line sensors, a start signal, a 15-second interval signal, and starting-line violation signals for each side of the ramp. The judge should have a button that initiates the start sequence. A series of periodic beeps that mimic the words, "ready, set, go!" should sound, with the final "go" being a loud and clearly distinguishable tone or buzzer. In addition, a green light or LED should illuminate when the "go" signal is sounded. The system should time for 15 seconds, then sound another tone or buzzer to indicate the end of the 15-second time interval. If a vehicle crosses the starting line prior to the "go" signal, a red light should go on for the violating vehicle's side of the ramp, and a special "violation" signal should be sounded to alert the judge.

39. Teams in the Peak-Performance Design Competition are called to the floor when it is their turn to compete. After the initial call, each team has three minutes to arrive at its starting line. A team that does not show up at the starting line after three minutes loses that run. Warnings are supposed to be given two minutes and one minute before the deadline. Traditionally, the announcer has called the teams and issued these warnings orally. With three races running simultaneously, all starting at different times, and with only some teams requiring the full three minutes to arrive, the proper issuing of these cues has been lax. The judges have asked that you design an automated system that will inform a given team how much of its three-minute sequence has elapsed. The system must send an appropriate signal, oral, auditory, or visual, to only the affected team, and the timing sequence must be initiated from the judges bench. As many as eighty teams may compete on a given day, and each is assigned one work table from a large array of 3×8-ft. tables where the competition is held. Typically, the team being called is delayed, because it is repairing or modifying its vehicle.

    One of the key design issues is whether a wireless or wired system is the better system, given the logistic constraints of the competition environment. As the event operates on a strict budget, ultimate cost is also an important factor.

40. Vehicles entering the Peak-Performance Design Competition must meet several constraints, including a maximum battery-voltage limit. Each vehicle is checked once with a voltmeter at the start of the day by the head judge. Having a standardized voltage-checking device would shorten the time for voltage checking. Design a unit that has a rotary (or other type of)

switch that can select a predetermined battery voltage. If the measured battery falls within the acceptable range, a green light should come on. If the voltage falls below or above the range, yellow and red lights, respectively, should come on.

41. A local company employs several workers who sort and package small (1 to 2 cm) parts in the 10–100 gm range. A typical operation might consist of putting ten small parts in a polyethylene bag for subsequent packaging. As part of a course design project, your professor has asked you to design a mechanical sorting apparatus for dispensing these parts one at time so that the employees do not have to pick them up by hand. Develop an outline for how such a system might work, and draw a sketch of your proposed apparatus.

42. Ace Cleaning Services employs custodians performing a variety of cleaning tasks at four major building in the downtown area. Many of these individuals have poor cognitive abilities and are unable to generalize the cleaning skills they have been taught. They cannot perform a supervisor-demonstrated office-cleaning task in a different office, even if a similar operation is involved. As a result, their cleaning job performance is often poor. Many of the employees rely on supervisor cues to repeatedly perform to acceptable standards the same task in different locations.

   A crew from Ace has been assigned to a 22-story office building with over 500,000 sq. ft. of space that requires cleaning. Approximately thirty employees service this building between 6:00 am and 10:00 pm with a staffing pattern of nine people. Workers are dispersed throughout the building to perform their daily cleaning routines. Supervisors are responsible for training and for checking that the work has been performed to acceptable standards. Many of the supervisors also perform direct labor and hence cannot consistently provide prompting or cues to the rest of the employees all the time. What is needed is a communication system that can provide on-the-job cues to the cleaning staff. One system is needed for individuals who are literate and another for individuals who are not. Your task in this project is to design a modular system of either type. The system must be designed so that additional units can be reproduced at low cost. As part of a project in your engineering design class, your professor has asked you to devise a system for providing recorded cues to Ace's employees. Develop a concept for the system, draw a sketch of one implementation, and outline how such a system might work.

43. A teacher at the nearby Carver School teaches a student who has severe developmental delays. This student is highly motivated by the "Wheel of Fortune™" TV game show. The teacher currently has a four-foot-diameter, wall-mounted, colorfully painted cardboard circle that spins and simulates the real game. Use of this wheel is supplemented by video clips of the game played on a VCR. The teacher would like a more elaborate, electronically interfaced version of the game that enables the spinning dial to activate lights, voice, and the VCR clips. Despite the circus-like nature of this project, it is a top priority of the Carver school system. The customer is in need of an imaginative and creative response to this problem, and your engineering professor has asked you for ideas. Sketch several versions of the system, highlighting the advantages and disadvantages of each approach. How would you test the success of your design?

44. A teacher wants a clock system that can help students to learn the relationship between time displayed by digital clocks and time displayed by analog clocks. The system should have a console that contains a large analog clock face, as well as a digital clock display with large digits. In operation, the teacher will set either clock, then ask the student to set the other clock to the same time. If the student sets the time correctly, the unit should signal the student appropriately. If the student fails to set the time correctly, the unit should also issue an appropriate response. Outline the salient features of such a system.

45. Your school has been asked by an individual confined to a wheelchair to build a small motorized flagpole that can be raised and lowered by pressing buttons. The person needs such a device to hold a bright orange flag to provide visibility outdoors while navigating busy city streets and sidewalks. The flag must be lowered when the individual enters buildings so that the pole does not interfere with doorways and low ceilings. Here is a copy of the letter received by your school:

East Crescent Residence Facility
11 Hastings Drive
West Walworth, MA 02100

Prof. Hugo Gomez
College of Engineering
Canton University
44 Cummington St.
Canton, MA 02215

RE: Retractable Flag for Wheelchair

Dear Professor Gomez,

> I am the supervisor of a residence facility that services adults with special needs. One of our residents is confined to a wheelchair and spends a great deal of time traveling throughout the community, often on busy streets, in a motorized wheelchair. Although a flag on a long pole would increase her safety, she is reluctant to install one on her wheelchair, because it becomes a problem in restaurants, crowded stores, and on public buses. I was wondering if you might have some students who could design an electrically retractable flag, possibly with visual enhancement (e.g., a flashing light) that could be raised or lowered by the individual on demand. The flag deployment mechanism could operate from either the wheelchair's existing automobile-type storage battery or its own self-contained battery. If such a device is possible, could you give me a call? I would appreciate any assistance that you might be able to offer.

Sincerely yours,

Liz DeWalt
Director

Prof. Gomez has asked you to try to build such a device for Ms. DeWalt. How would you approach such a task? Such a seemingly simple device actually can be more complicated to design than you might think. Draw a sketch of the wheelchair device, then devise a design plan for building and testing several designs. Include a list of possible safety hazards to bystanders and the user. How can you include the user in the design process, and why is it advisable to do so? Also write a report of your preliminary findings for Prof. Gomez. Your design strategy should begin with a conceptional drawing of the device that you can send to Ms. DeWalt for comments. Generate a specification list and general drawing of the apparatus as well as a cover letter to Ms. DeWalt.

Design Considerations   One goal of your design might be to make a flag device that can be mounted on *any* wheelchair, not just on that of Ms. DeWalt's client. Because many wheelchairs are custom designed for the user, your device must be easily adaptable for mounting on different wheelchair styles. Not all wheelchairs are motorized, hence your device must operate from its own batteries to accommodate hand-pushed wheelchairs. Another consideration in favor of separate battery power is that motorized wheelchair manufacturers usually specify that no other electrical or electronic equipment be connected to the primary motor battery for reasons of safety, reliability, and power integrity.
  One last consideration concerns the placement of the switch needed to activate the flag. Like the flag itself, the activation switch must easily attach to structural features of the wheelchair and must be within easy reach of the user. At the same time, it must not be so obtrusive that it distracts the user when not in use and must not hurt anyone.

46. Develop a design concept for a computer-interfaced electronic display board that can be placed in the lobby of an office building to display messages of the day, announce upcoming seminars, or indicate the location of special events. The objective of the problem is to use a

matrix of addressable light-emitting diodes (LEDs) rather than a video display. The system should accept messages by wire from a remote site. One approach might be to design your display board system so that it is capable of independently connecting to a local area network via a telephone line. Alternatively, you could build a separate remote device that could be connected to a desktop computer and then brought down to the display board to load in the data. These examples are suggestions only. In general, any means for getting data to the board is acceptable, but a separate computer cannot become a dedicated part of the finished display.

47.   An engineer is interested in measuring the small-valued ac magnetic fields generated by power lines and appliances. You have been asked to design a battery-powered, hand-held instrument capable of measuring the magnitude of ac magnetic fields in the range 0.1 to 10 μT (microtesla) at frequencies of 50–60 Hz. Magnetic fields of this magnitude are very small and are difficult to measure accurately. For comparison, the earth's dc magnetic field is on the order of 50 μT, and the magnetic field inside a typical electric motor is on the order of 1 T.

Using your knowledge of physics, summarize the important features that such a device should have. Outline a design plan for its development and construction. You have several options for the primary sensor. For example, it may consist of a flat coil of wire of appropriate diameter and number of turns, or, alternatively, you might consider using a commercially available semiconductor sensor. Note that dc fields, such as those produced by the earth or any nearby permanent magnets, are not of interest. Hence, any signal produced by dc fields in your instrument should be filtered out. Ideally, your unit should have a digital or analog display device and should accommodate a remote probe if possible.

48.   A friend of yours runs a residence home for individuals with mental and physical handicaps. She would like a medicine dispenser for dispensing pills at specific times. The unit is to be carried by an individual and must have sufficient capacity to hold medication for at least one day. The unit should open a cassette or compartment and should emit an audible or visual signal when it dispenses medication. The unit must be easy to load and should be easily programmable by a residence-home supervisor. Your friend has asked you to assess the feasibility of developing such a unit. Get together with up to four other students. Discuss the feasibility of the idea and develop one or more design concepts for implementation.

49.   A friend or yours is an enthusiast of remote-control model airplanes. One perennial problem with radio-controlled airplanes concerns the lack of knowledge about the flight direction and orientation of the airplane when it is far from the ground-based operator. When the airplane is too far away to be seen clearly, the operator loses the ability to correctly control its motion. Develop a design concept for a roll-, pitch-, and compass-heading indicator system that can be mounted on the model airplane and used to send the information via radio back to the operator's control console. Your system should sense the pitch and roll of the airplane over the range +90 degrees to −90 degrees and be able to withstand a full 360-degree roll or loop-de-loop.

50.   Your family has asked you to design a remote readout system for a vacation home to be interrogated by a remote computer over a modem and telephone line. The unit in the vacation home should answer the phone after ten rings, provide means for an entry password, and then provide the following information:

    Inside and outside temperatures,

    Presence of any running water in the house,

    Presence of any loud noises or unusual motion,

    Status of alarm switches installed on doors and windows.

Discuss the design specifications for such a unit and develop a block diagram design for its implementation.

Engineering Ethics   The following problems on ethics have no clear answers. They are included in this problem section to stimulate discussion on engineering ethics.

51. No mass-produced product can be made defect free 100 percent of the time. Suppose that you have the job of designing the gasoline tank for a new model of automobile. One of the problems with gas tanks is that they can rupture upon impact, leading to fire and explosion. You've determined that the cost of adding a metal baffle shield behind the gas tank to protect it in the event of a rear-end collision is approximately $50 per vehicle (including labor). The shield will add about 10 kg of weight to the vehicle, reducing its gas mileage by an average of about 0.2 percent. Your statistical models suggest that 0.1 percent of the cars sold will be involved in rear-end collisions during their product lifetime. Given that approximately 100,000 of the cars will be sold in the first model year, should you advocate for adding the protective baffle, or should you leave it out to save money and increase profit margin?

52. You have worked for eight years for Hobo Electronics designing microchips. During the past year, you helped design a state-of-the art processor that significantly surpasses the performance of your competitors' microchips. This past year also has been difficult financially for Hobo, and you are given a layoff notice. You begin searching for a new job and find a strong lead at Beta Corp as a microchip designer. During the interview, your potential boss makes it clear that you can be hired at a substantial salary increase if you are willing to bring along with you the design concepts from Hobo's new chip. Should you take the job, or do you have a professional obligation to Hobo keep their proprietary information secret?

53. You've answered an ad for a part-time job as a spreadsheet programmer. You are extremely familiar with Microsoft Excel®, and figure that you can handle the job. During the interview, you learn that your potential employer really wants someone who can code spreadsheets in Lotus 1-2-3®. Do you pretend that you know Lotus 1-2-3® because you assume it's probably similar to Excel and that you can learn it very quickly? Or do you tell the truth about your actual experience and possibly risk not being offered the job?

54. You work for a company that produces microprocessor-based systems. A customer has asked for a control module with a specific set of features. You have adapted a program from a previous job for another customer. The borrowed module meets your customer's requirements, but it also includes features that the customer did not request. You do not mention these additional features to the customer when you deliver the product.

  A few months later, the customer calls you to express satisfaction with the product and requests an updated version that includes additional features. These newly requested features are identical to the extraneous features from the original borrowed module. Do you accept the job, do virtually no additional work, and charge the customer again for the "added" features, or do you tell the customer that his present module already has the requested features?

55. In reviewing a recent exam, you notice that your professor has failed to see one of your errors and has given you too much credit for the problem. You know that you wouldn't hesitate to mention a grading error if you had been given too few points. Do you call the oversite to the attention of your professor, or do you say nothing and accept the extra points?

56. Your employer has asked you to dump a drum filled with cleaning solvent on the ground out in back of your company's manufacturing facility. You know that the town water supply is nearby, a distance of about two hundred yards. You raise your concern with your immediate boss who tells you that the company has been dumping small amounts of the chemical in the same spot for years and that no one has complained about it yet. "It's only a small amount," your boss says, "We certainly wouldn't dump any large amounts, because that would contaminate the water supply. It would cost us a fortune to have only one drum carted away by a toxic waste disposal company, though. I'm sure the solvent evaporates long before it can ever get to the water supply." Do you refuse to comply with your employer's request, or do you dump the cleaning solvent?

57. You work for a company that produces hand tools. You have designed a steel and acrylic framing hammer used to insert and remove nails. Several from the most recent lot of 20,000 have been returned from contractors with broken handles, and one caused injury when its handle broke while striking a nail. You've traced the problem to a contaminant in the acrylic resin used to make the handles and note in the production log that the problem was corrected

after the first 200 hammers. Do you suggest to your supervisor that the company issue a recall order for all 20,000 hammers, offer a free replacement for any that are returned with broken handles, or do nothing and wait to see how many are returned?

58. You've recently been given a promotion and salary increase, but a colleague and rival at work has not. You learn that your rival is making false statements about your work to your immediate superior. In particular, he claims that you falsified data on a recent benchmark test so that you could get your product out the door ahead of schedule. You discuss this problem with your supervisor who tells you to resolve the conflict with your jealous rival on your own, because if word of the discord finds its way to the company president, then your supervisor will look bad. Weeks go by, and nothing changes. Do you do nothing and assume that your reputation will win out over the false statements of your rival, do you take your problem to the president, or do you look for another job?

3

Engineering Design Tools

The engineer utilizes many technical skills when designing a device, product, or system. Prototyping, breadboarding, estimation, modeling, simulation, computer analysis, and testing are all part of the engineer's tool kit. An engineer also must have numerous nontechnical skills, including the ability to work in a team, generate new ideas, make sketches, keep good documentation, make approximations, and predict outcomes. Seldom does an engineer simply sit down and get right to work on the technical details of a project. Before beginning work in earnest, he or she spends much time planning, conducting feasibility studies, reviewing results of other projects, doing approximate calculations, interacting with other engineers, and defining the approach to the problem. This chapter introduces several of the skills—both technical and nontechnical—that an engineer uses as part of the design process.

SECTIONS

- 3.1 Teamwork as a Design Tool
- 3.2 Brainstorming
- 3.3 Documentation: The Importance of Keeping Careful Records
- 3.4 Estimation as an Engineering Design Tool
- 3.5 Prototyping and Breadboarding: Common Design Tools
- 3.6 Reverse Engineering
- 3.7 Project Management

OBJECTIVES

In this chapter, you will learn about

- Design tools available to the engineer.
- Brainstorming, documentation, estimation, prototyping, and project management.
- Establishing foundations for good engineering design.

3.1 TEAMWORK AS A DESIGN TOOL

The spirit of rugged individualism persists as a theme in books, movies, and television. The image of a lone hero or heroine who strives for truth and justice against insurmountable odds may appeal to our sense of adventure and daring. The dream of becoming a sole entrepreneur who endures economic and technical hardship, eventually prevailing with an award-winning product leading to riches, arouses our pioneering spirit. Yet, in the real world, engineers seldom work alone. Most engineering problems are interdisciplinary, and true progress requires teamwork, cooperation, and the contributions of many individuals. This concept is easy to understand in the context of designing large structures, such as bridges and buildings, or world-wide computer networks. Likewise, complicated devices, such as automobiles, video players, medical implants, network routers, copy machines, and ink-jet printers, cannot be designed by one person alone. Some of the great accomplishments in space exploration of the 20th century, such as the Apollo moon landing, the Mir space station, and the Hubble telescope, required hundreds, if not thousands, of engineers working in cooperation with each other and teams of physicists, chemists, astronomers, material scientists, medical specialists, and mathematicians. Teamwork is an important skill, and you should learn it as part of your engineering education. (See Figure 3.1) Working in a team requires that you speak clearly, write efficiently, and have acquired the ability to see another person's point of view. Each member of a team must understand how his or her task relates to the responsibilities of the team as a whole.

Many engineering firms offer team building workshops as part of their employee training programs, and self-help books on the subject of teamwork abound. You'll have many opportunities to work as a team if you study engineering, and you should treat each one as a learning experience.

Effective Team Building

An effective team is one that works well together and functions at its maximum potential when solving a design problem. One key ingredient of an effective team is a good attitude toward fellow teammates and team activities. Team morale and a sense of pro-

Figure 3.1. Students working on a design project build strong team relationships.

fessionalism can be enhanced if team members agree upon a set of rules of behavior. The following set offers one possible guideline for effective team building.

1. Define Clear Roles

Each team member should understand his or her function within the organization of the team. The responsibilities of each individual should be defined *before* work begins on the project. Roles need not be mutually exclusive, but they should be defined so that all aspects of the design problem fall within the jurisdiction of at least one person. In that way, no task will "fall between the cracks."

2. Agree Upon Goals

Members of the team should agree upon the goals of the project. This consensus is not as easily achieved as you might think. One teammate may want to solve the problem using a traditional, time-tested approach, while another may want to attempt a far out, esoteric path to success. Define a realistic set of goals at the outset. If the design process brings surprises, you can always redefine your goals midway through the project.

3. Define Processes and Procedures

Teammates should agree on a set of procedures for getting things done. Everything from documentation and the ordering of parts to communication with professors, clients, and customers should follow a predetermined procedure. In that way, misunderstandings about conduct can be greatly reduced.

4. Develop Effective Interpersonal Relationships

You must learn to work with everyone on your team, even those individuals who you may personally dislike. In the real world, a client will seldom care about what conflicts may occur behind the scenes. It's a sign of engineering professionalism to be able to rise above squabbles and personality clashes as you concentrate on the job at hand. Be nice. Be professional. Forbid name calling, accusations, and assigning fault between team members.

5. Define Leadership Roles

Sometimes a team works best when a single person emerges as a clearly defined leader. Other teams work better by consensus using distributed leadership, or even no leadership at all. Regardless of your team's style, make sure that leadership roles are clearly defined and agreed upon at the start of a project.

PROFESSIONAL SUCCESS: YOU'RE THE TEAM LEADER, ONE TEAMMATE HAS DISAPPEARED, AND YOU'RE DOING ALL THE WORK

It's impossible to get along with all people all the time. When you work in close proximity to other individuals such as project teammates, personal conflicts and disagreements are inevitable. At times these personal differences occur because one team member fails to meet his or her responsibilities. Remember that however complicated your team relationships may become, your customer does not care about them. Your customer is interested in receiving a well designed product that reflects your best engineering abilities, so it's up to you to resolve team conflicts internally. At times this resolution will simply mean that you do more work than others, even if you will not be rewarded for your extra efforts. A good leader understands this tradeoff and devises a plan to work around the errant teammate. Such situations may seem frustrating and unfair, but they happen in the real engineering world all the time. Learning how to deal with them as a student is part of your engineering training.

Exercises 3.1

E1. Show how the five elements of effective team building might apply to the functioning of a basketball team.

E2. Define the roles, goals, and procedures that might apply to the design and construction of a suspension bridge.

E3. Define the roles, goals, and procedures that might apply to a team of software engineers developing a Web site for the sales catalog of a national book selling chain.

E4. Define the roles, goals, and procedures that might apply to a team of electrical and mechanical engineers developing an electric automobile.

E5. Define the roles, goals, and procedures that might apply to a team of biomedical engineers developing an artificial heart/lung machine.

3.2 BRAINSTORMING

One obvious area in which teamwork plays an important role is the generation of ideas for problem solving. When engineers gather to solve problems, they often resort to a creative process called *brainstorming*. Brainstorming requires a spontaneous mode of thinking that frees the mind from traditional boundaries. All too often, we limit our problem solving approach to obvious solutions that have worked in the past. Responsible engineering sometimes requires that we consider other design alternatives, including those previously untried. A good engineer will never settle on a solution just because it's the first one to come to mind. When engineers brainstorm, creativity proceeds spontaneously unfettered by concerns that an idea is "way out" or impractical. Hearing the ideas of others taps new ideas buried in the recesses of the brain. Ideas are discarded as unfeasible only after consideration, study, analysis, and comparison with competing ideas. Brainstorming allows the engineer to consider as many options as possible before choosing the final design path.

Brainstorming can be done informally, or it can follow one of several time-tested formal methods (See Figure 3.2). Formal methods are used in large group settings where organization is needed to avoid chaos and anarchy. Informal methods typically are used when one, two, or perhaps three people wish to generate ideas. Although they differ in execution, informal and formal brainstorming techniques share the same set of core principles. The primary goal is to foster the inhibited, free exchange of ideas by creating a friendly, nonjudgmental environment. Brainstorming is an art and requires practice, but anyone who has an open mind and some imagination can do it.

Ground Rules for Brainstorming

The ground rules for brainstorming are designed to create a friendly, nonthreatening environment that encourages the free flow of ideas. Although the specific rules may vary, depending on the procedures followed, the following list can serve as a guideline:

1. No holding back. Any idea may be brought to the floor at any time.
2. No boundaries. An idea is never too outrageous or "way out" to mention.
3. No criticizing. An idea may not be criticized until the final discussion phase.
4. No dismissing. An idea may not be discounted until after group discussion.
5. No limit. There is no such thing as having too many ideas: the more the better.
6. No restrictions. Participants may generate ideas from any field of expertise.
7. No shame. A participant should never feel embarrassed about bringing up a stupid idea.

Figure 3.2.   Students engage in an idea trigger session.

When a brainstorming session is in progress, one person should act as the facilitator, and another should record everyone's ideas. It's also possible to use video or audio tape in lieu of a human secretary.

Formal Brainstorming Method

When a large group gets together to brainstorm, some sort of formal structure is needed. Without such a structure, a flood of competing ideas, all brought to the floor simultaneously, can create chaos. Instead of thinking creatively, participants become confrontational as they strive to be heard and gain a voice in the conversation. With so many randomly competing opinions, each person's creative process is inhibited, and the brainstorming session becomes counterproductive. This effect is sometimes called "idea chaos." Adding formal structure to the brainstorming session restricts the flow of ideas to a manageable rate without restricting the number of ideas generated. In fact, the addition of formal structure can enhance the brain's creative process in large group settings by preventing aggressive individuals from dominating the conversation and by providing time for people to think.

Of the many formal brainstorming techniques that exist, the *idea trigger* method has been well tested and is used often by brainstorming specialists in large group settings. The idea trigger method is based on the work of psychologists[*] and has been shown to enhance the brain's creative process. It may seem contrived at first, but it's been well tested and has been shown to be effective. It relies on a process of alternating tension and relaxation that taps the brain's creative potential. By listening to the ideas of others, receiving the foreign stimulus of other people's spoken ideas, and being forced

[*]G. H. Muller, *The Idea Trigger Session Primer,* Ann Arbor, MI: A.I.R. Foundation, 1973. S. F. Love, *Mastery and Management of Time,* Englewood Cliffs, NJ: Prentice-Hall, 1981.

to respond with counter ideas, a participant's habitual behavior patterns, personality traits, and narrow modes of thinking, which often serve as barriers that stifle creativity, can momentarily be broken, allowing ideas hidden in the recesses of the brain to come to the foreground. A participant who is shy, for example, and reluctant to offer seemingly silly ideas will be more willing to do so under the alternating tension and relaxation of the purge-trigger sequence.

The idea trigger method requires a leader, at least four participants, and a printed form, such as the one shown in Figure 3.3. The procedure has three phases, as follows.°°

Phase 1: Idea Purge Phase   The problem or design issue is summarized by the leader. Each person is given a blank copy of the form shown in Figure 3.3. Without talking, each participant writes down in rapid succession as many ideas or solutions as possible. These entries are placed under Column 1. Key words suffice; whole sentences are not necessary. During the idea generation phase, participants open their minds, consider many alternatives, and do not worry if ideas seem too trivial or ridiculous. "Pie in the sky" types of ideas that may seem radical or impossible should also be included. In short, participants write down *anything* that comes to mind that may be relevant to the

Figure 3.3.   Blank form for the idea trigger session.

IDEA TRIGGER SESSION

COLUMN 1	COLUMN 2	COLUMN 3	COLUMN 4
120 MINUTES			
60 SECONDS			

CONTRIBUTOR:

°°C. Lovas, *Integrating Design Into the Engineering Curriculum,* Dallas, TX: Engineering Design Services Short Course and Workshop, October 1995.

problem. The fact that ideas are written down silently removes the element of intimidation from the idea generation process.

After the first two minutes of the session, the group takes a break and then attempts to write down additional ideas under Column 1 for another sixty seconds. This *tension and relaxation* sequence has been shown to enhance creativity. It helps to completely extract all ideas from the brain's subconscious memory, much like squeezing and releasing a sponge several times to extract all the water.

Phase 2: Idea Trigger Phase   After the idea generation phase, the leader calls upon all members of the group. Each participant takes a turn reading his or her entries from Column 1. As each person recites Column-1 entries, others silently cross out the duplicates on their own lists. Hearing the ideas of others will trigger new ideas, which each person should enter under Column 2 as soon as they emerge. This process is called *idea triggering*. Hearing the remarks and ideas of others while pausing from the act of speaking causes the hidden thoughts stored in the subconscious to surface. The purpose of the idea trigger phase is not to discount the ideas from Column 1, but rather to amplify them, modify them, or generate new ideas.

After all members have read their Column-1 entries and have completed their Column-2 entries, the idea trigger process is repeated again. This time, entries from Column 2 are read, and any new ideas triggered are entered under Column 3. The process is repeated, with entries added to Columns 4, 5, etc., until *all* ideas are exhausted. Complex problems may require as many as five rounds of idea trigger phase.

The entries that appear under the second and third columns (and the fourth and fifth columns if the problem is complex) are usually the most creative. Such richness is thought to result from several factors. Often participants are secretly angered at having had their ideas stolen by another and are self-motivated to move on to new, unexplored territory. Simple competitive pressure can also propel a person toward new, original ideas. Conversely, seeing that one's ideas have not been duplicated by others can provide positive reenforcement, pushing the participant to come up with even better or more refined ideas. Some individuals may respond to their own nonduplicated entries with a desire to produce more as a way of hoarding the good ideas. Yet others may subconsciously think that augmenting previously discussed ideas fosters group cooperation.

Phase 3: Compilation Phase   When the idea trigger phase has been completed, it's the job of the leader to compile everyone's sheets and make one master list of all the ideas that have been generated. The group then proceeds to discuss all ideas, discarding the ones that probably will not work, and deciding which of the remaining ideas are appropriate for further consideration and development.

An Example of Formal Brainstorming

Let's illustrate the formal idea trigger method with an example. Four students, Tina, Juan, Fred, and Karin, are designing an entry for the Peak-Performance Design Competition introduced in Chapter 2. As discussed in that chapter, the overall objective is to design a self-propelled vehicle that can climb a 1.5-meter ramp, stop at the top, and prevail over an opposing vehicle climbing up the ramp from the other side. The four students recently held a formal brainstorming session based on the idea trigger method. They agreed to address all elements of the car design, including the issues of propulsion, offensive and defensive strategies, and the stopping mechanism. The following discussion chronicles their brainstorming session. Tina acted as the leader and timed the first two minutes, the break, and the subsequent 60-second idea purge phase. At the end of the purge phase, Juan's page looked like this:

JUAN	BRAIN PURGE PHASE COLUMN 1 2 MINUTES:
	• Support structure = wood (easy to make)
	• Use angle irons from Mechano™ (Erector™ Set)
	• Plastic body for lighter weight
	• Zinc air batteries (lightweight)
	• Wheels taken from my radio controlled car
	• Rubber band for chain drive
	• Small car will be harder for opponent to deflect.
	1 MINUTE:
	• Ramming device
	• Wedge shaped body

Juan read his entries. As Fred listened to Juan, he crossed out his own duplicate entries. When Juan was finished, Fred's first column, including crossouts, looked like this:

FRED	BRAIN PURGE PHASE COLUMN 1 2 MINUTES:
	• ~~No heavy batteries (use zinc air)~~
	• Larger wheels for slower turning speed
	• Gear box
	• Higher torque (harder for opponent to push backwards)
	• ~~Use plastic for body~~
	• Electronic timer for stopping mechanism
	• Rechargeable batteries
	• ~~Wedge shaped design~~
	1 MINUTE:
	• ~~Buy wheels from hobby shop for radio controlled car~~
	• Sense speed, determine distance traveled
	• Aluminum frame

Karin next read those of her entries that had not been duplicated by Juan. As Fred listened to Karin, an idea flashed into his head. *A threaded rod,* he thought, *We can make the drive shaft from a threaded rod.* Fred reasoned that they could make the drive shaft from a threaded rod and have it screw a sliding nut toward a cutoff switch. The method would not be foolproof, because slipping wheels could ruin the system's ability to track distance, but it was worth discussing. Fred wrote down "threaded rod" under his Column 2 entries.

When Karin heard Juan read his "ramming device" entry, it had made her think about using an ejected object as part of an offensive strategy. She wrote the words "ejected device" under her Column 2 entries. Tina reacted similarly to Juan's idea and wrote the words, "lob something on the track ahead of opposing car" under her Column 2 entries.

The spoken trigger phase made its way around the group. A great many ideas, some simple, some esoteric, and some very clever found their way onto peoples lists.

When everyone had finished, Tina started the process again. This time, everyone read their Column 2 entries and wrote down new ideas under Column 3. As Karin read her entry about ejected devices from Column 2, Tina got another idea. The idea of an ejected object brought to her a fleeting image from the Herman Melville novel *Moby Dick.* She imagined a flying spear with a barbed tip shot ahead of the vehicle over the top of the hill. *After hitting the carpet in front of the opposing vehicle,* she thought, *the barbed tip will dig into the carpet, blocking the other car, and be very difficult to dislodge.* Tina wrote down "harpoon" under her Column 3 entries.

The second idea trigger round progressed, and Tina started yet another one. After about 45 minutes, the entire Phase 2 session was finished. Tina suggested a break so that she could compile everyone's lists of ideas. Her combined list of entries from everyone's three columns looked like this:

COMPLETE LIST OF IDEAS FROM EVERYONE'S SHEETS

SHAPE:

- Small car = harder for opponent to deflect
- Wedge shaped vehicle having same width as track
- Rolling can design
- Snow plow shaped wedge

STRUCTURE:

- Support structure = wood (easy to make)
- Aluminum frame
- Plastic body for light weight.
- Use angle irons from Mechano™ (Erector™ Set)
- Hot melt glue balsa wood

POWER:

- Zinc air batteries (lightweight)
- Rechargeable batteries
- Change batteries after every run
- Electronic timer for stopping mechanism
- Microprocessor-controlled car with onboard sensors
- Sense speed, determine distance traveled from microprocessor software

PROPULSION:

- Wheels from radio-controlled car purchased at hobby shop
- Large wheels
- Rubber band for chain drive
- Plastic linked chain from junked radio-controlled car chassis.
- Single large mousetrap with mechanical links
- Wind up large rubber band

STRATEGIES:

- Ramming device
- Flying barbed harpoon
- Pick up arm
- Throw jacks in front of oncoming opponent
- Roll over opponent with large roller

After the break, Tina reconvened the team to discuss the list of ideas. They weeded out the ones that did not seem feasible and compared ideas that looked promising. They combined multiple ideas and converged on a slow-moving, wedge-shaped vehicle concept for the prototype stage. They also decided to try out Tina's offensive strategy of a flying harpoon designed to dig into the carpet and block the path of the opposing vehicle.

Informal Brainstorming

The formal brainstorming method discussed in the previous section requires organization and planning. In contrast, informal brainstorming can be done anywhere. Have you ever had a thinking session with a fellow student, friend, or colleague? If so, you probably engaged in informal brainstorming. You need not have discussed a technical issue. The conversation might have revolved around something as mundane as a social get-together. Suppose, for example, that you tried to figure out transportation arrangements for a trip to a show at the science museum. The conversation might have proceeded as follows:

You: "I've called the Science Museum, and the 11- and 1-o'clock shows are sold out. We can get seats for 3, 5, and 7 o'clock. What do you think?"

Friend: "Okay, it's 10 o'clock now. If we aim for the 5-o'clock show, we can head for dinner afterwards."

You: "How about asking Pat to join us?"

Friend: "Great!"

You to friend (after phoning Pat): "Well, Pat wants to come but has another appointment at six. Will that give us enough time to see the 5-o'clock show? "

Friend: "Not really."

You: "How about this: We eat an early lunch, go with Pat to the three-o'clock show, and see as much of the general exhibits as we can before Pat has to leave? We can stay on until the Museum closes at nine if we like."

Friend: "Sure, sounds fine. Give Pat a call."

You (with Pat on hold): "It's okay with Pat. What time is the next bus?"

Friend: "Ten thirty. But, if we take the commuter train, we can meet Pat at the museum and eat at the museum cafeteria."

You: "Yes, but the food's not great there. Too healthy!"

Friend: "How about if we stop at Mo and Jo's Submarine Palace, buy sandwiches for three, and eat them at the museum?"

You: "Better yet, we can have a picnic on the lawn in back."

Friend: "Good idea. Ask Pat if that idea works."

You and your friends engaged in informal brainstorming over a very ordinary, everyday topic: going to the museum. This commonplace discussion had in it the same features that one would find in a technical brainstorming session. The problem had many solutions, and arriving at the final choice involved iteration and testing. By trading thoughts, bringing several ideas to the table, discarding those that were judged not feasible, and building upon those that were, you eventually arrived at a plan of action.

Informal Brainstorming: An Engineering Example

The informal brainstorming technique can be used to solve technical problems that arise during the design process. As a technique for engineering design, informal brainstorming in a round table format is appropriate for small groups of people. Ideas are contributed in random order by any participant. The flow of ideas need not be logical, and new proposals can be offered whenever they come to mind.

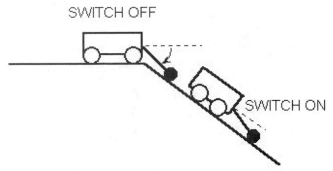

Figure 3.4. Juan's idea for a stopping switch.

The following conversation provides an example of an informal brainstorming session involving a technical topic. It chronicles again an imaginary discussion that took place between two students attempting to design an entry for the Peak-Performance Design Competition introduced in Chapter 2. Two of the students from the previous example, Tina and Juan, discuss the issue of how to make the vehicle stop when it arrives at the top of the ramp. Take note of the flow of ideas and the way in which the design concept evolves with the conversation. Tina and Juan do not settle on the first idea that comes to mind, but instead allow the flow of ideas to lead them to better solution.

Tina: "I think we can use a switch of some kind to turn off the electric motor when the car arrives at the top."

Juan: "Yes, that's one possibility." (*He thought for a while*) "We also could modify the switch so that it trails behind and is spring loaded in the closed position. Putting the car on a flat surface will press the lever arm against the spring, closing the switch and connecting the battery to the motor. When the car reaches the top, the switch arm will stay on the slope, allowing the spring to force the arm downward. The switch will open, disconnecting the battery from the motor and stopping the car." (*Juan drew the sketch of Figure 3.4.*)

Tina: "It *might* work, but I'm afraid that the switch might open before the car gets to the top of the ramp. Unless we measure and construct everything perfectly, the switch may open prematurely when the car just starts to go over the transition to the top of the hill." (*She drew the sketch shown in Figure 3.5.*)

Juan: "Yes, premature switch opening may be a problem. We could try it out."

Tina: "Here's another idea: We could connect a long screw thread to the drive shaft of the car and put some sort of sliding nut along the shaft. As the shaft turns, it will move the nut toward a stationary, normally closed switch, opening it after just the right

Figure 3.5. Premature opening of the switch.

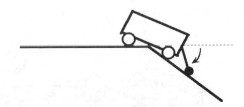

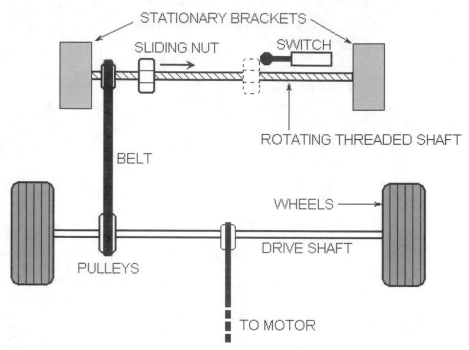

Figure 3.6. Non-turning nut moves along a rotating, threaded shaft.

amount of distance traveled. It might look something like this:" (*She drew the sketch shown in Figure 3.6 and presented it to Juan.*)

 Juan: "Yes! We could self-calibrate it before each run by placing the car on the top of the hill with the nut in the 'off' position against the switch and then manually roll the car *backwards* to the starting point. As the nut traveled back along the threaded rod, it would arrive in just the right position for the start." (*He thought a moment more.*) "How about a butterfly wing nut on the rotating shaft? We could buy one at the hardware store instead of having to make our own. One wing of the nut could ride against the car frame and prevent the nut from turning as the shaft turns. In this way, the nut would be screwed down along the threaded shaft. The other wing could be used to press the switch." (*Juan drew the sketch shown in Figure 3.7.*)

 Tina: "Yes, we could use one of those switches that has three contacts: normally open (NO), normally closed (NC), and common (COM)." (See Figure 3.8.) "The switch remains in the closed position, thereby connecting COM to NC, until it is pressed. When it's pressed, it switches COM to the NO contact."

Figure 3.7. The wing nut concept.

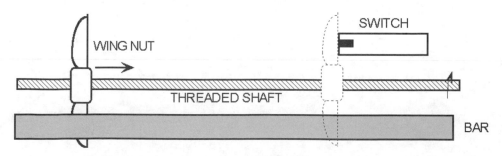

Figure 3.8. Switch showing normally open (NO) and normally closed (NC) contacts. The normal state of the switch refers to the condition where no force is applied to its pushbutton or lever arm.

After some thought, Tina continued, saying, "You know, we don't have to run a separate threaded rod from the drive shaft. We can make the actual drive shaft *be a* threaded rod, like this." (*Tina drew the sketch shown in Figure 3.9 and presented it to Juan.*)

"That way, we'll have a simpler design, reduce frictional loss, and provide an easy way to attach the wheels to the drive shaft with locking nuts."

Juan: "I like this new idea much better than our first switch idea. I think it's a lot more foolproof."

Tina: "On the other hand, if we use the threaded-rod idea and the wheels slip at all, the wing nut will still be threaded down the turning shaft, but the car won't be moving, and it will stop prematurely, short of the top of the hill."

Juan: "We could go back to our separate shaft idea and lubricate the shaft very well to reduce friction."

Tina: "No, that still won't solve the problem of the slipping wheels."

Juan: "Well, then, I have another idea. We could use a threaded shaft that is separate from the drive shaft and have it be turned by two idler wheels, rather than by the motor. Idler wheels will run along the track and turn the wing nut shaft but will be much less likely to slip because they won't be driven by the motor and thus will experience a much smaller torque." (*Juan drew the sketch shown in Figure 3.10.*)

"The idler wheels won't be connected in any way to the drive shaft or driven wheels. The only way that the idler wheels can turn is if the car is moving. If the idlers themselves get stuck, they'll drag along the track and *not* move the nut by the correct amount. But if the mechanism is well lubricated, stuck idler wheels should not be a problem."

Figure 3.9. Threaded rod also serves the function of a drive shaft.

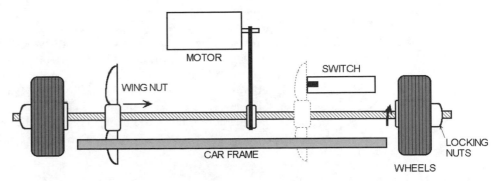

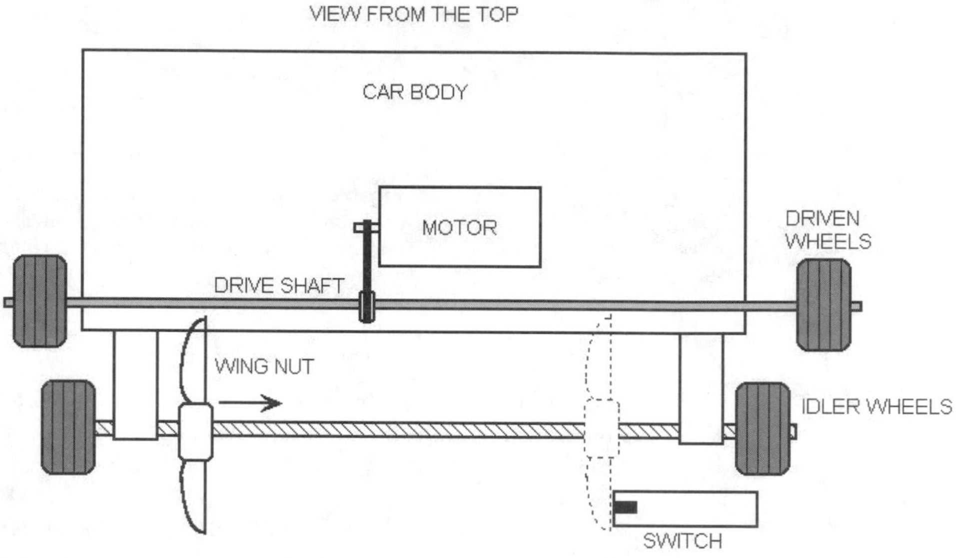

VIEW FROM THE TOP

Figure 3.10. Idler wheels turn the threaded rod which is not driven by the motor

Tina: "We've got some ideas to try, but I'd like us also to consider an electronic timer that keeps the motor turned on for a fixed amount of time. We could experiment with the car and determine the exact amount of time needed for it to get to the top."

Juan: "Or how about an altimeter that measures the height of the car off the floor?"

Tina: "Or maybe a string that hooks to the base of the ramp, unwinds as the car goes up, and pulls a switch to the off position when the car arrives at the top?"

Juan: "That's not allowed by the rules. All parts of the car have to lie inside the one-meter marks at the end of the fifteen-second time interval."

Tina checked the rules and confirmed that Juan was correct. They both considered the altimeter idea and decided that it was not feasible. *If* they could find one at all, an altimeter with a resolution in the one-meter range would be an expensive (and heavy) instrument indeed. Neither student wanted to design an altimeter from scratch. Their primary desire was to focus on the design of the vehicle itself. They also decided against the electronic timer idea for now. They were afraid that the car might travel at increasingly slower speeds from run to run as the battery ran down, and they remembered from their electronics class that the speed of the typical electronic timer is independent of its power supply voltage.

With their preliminary design concepts prioritized, they focused their attention on the two mechanical solutions: the trip switch and the concept of the sliding nut. They decided to build a few prototypes and try out these basic ideas.

The foregoing conversation between Tina and Juan illustrates the principles of informal brainstorming. As each person stated a new idea, the other amplified upon it and came up with new ones. In the end, the students condensed their list of ideas into one or two concepts that seemed feasible and agreed to try them out in test experiments. The flow of words between Tina and Juan was appropriate for two people and even would have worked for three or four. Had their design team been larger, a formal brainstorming method, such as the one discussed in the previous section, would have been more appropriate.

Suppose that you are the leader of a brainstorming session and one member of your team dominates the conversation. That person may criticize participants, dismiss unconventional ideas, cut off speakers, or otherwise break the rules. When this situation occurs, it's your responsibility to keep the offender in line. Say to the group, "Hey, we need to stick to the formal rules of brainstorming. I'm going to institute a don't-speak-until-called-upon rule." This approach will tactfully short circuit the behavior of the dominant participant while helping to maintain harmony among team members.

Exercises 3.2

E6. Conduct a one-person mini-brainstorm session and add as many ideas as you can to the final list of ideas compiled by Tina, Juan, Fred, and Karin. Allow yourself four minutes of brainstorm time to compile your ideas.

E7. A non-engineering friend complains about a pair of eyeglasses that keeps falling off. Give yourself five minutes of brainstorming time, and compile a list of as many ideas as you can for solving your friend's problem.

E8. Over a time span of two minutes, write down as many ways as you can for safely confining a dog to your back yard.

E9. Can brainstorming be used in solving math problems? Why or why not?

3.3 DOCUMENTATION: THE IMPORTANCE OF KEEPING CAREFUL RECORDS

Engineering design is never performed in isolation. As an engineer, it's your professional responsibility to record your ideas and the results of your work. One way that engineers communicate with each other is through careful record keeping at each stage of the design process. The collection of records, drawings, reports, schematics, and test results is referred to collectively as the *documentation trail.* The documentation trail serves as a tool for passing information on to individuals who may need to repeat or verify your work, manufacture your product from a prototype, apply for patents based on your inventions, or take over your job should you be promoted or move to another job. Written records are also a good way to communicate with yourself. Many an engineer has been unable to reproduce design accomplishments or confirm test results due to sloppy record keeping. Indeed, one of the marks of a professional engineer is the discipline necessary to keep accurate, neat, and up-to-date records. Documentation should never be performed as an afterthought. If a project is dropped by one engineer, the state of documentation should *always* be such that another engineer can resume the project without delay. As a student of engineering, you should learn the art of record keeping and develop good documentation habits early in your career. Most companies, laboratories, and other technical institutions require their employees to keep records that document the results of their engineering efforts.

Paper versus Electronic Documentation

Nowadays just about every piece of engineering documentation, with the exception of the engineer's notebook described in the next section, is generated on a computer.

Examples include word processing, spreadsheets, schematics, drawings, and simulated test results. All of this documentation must be preserved. Some engineers prefer to preserve documentation by printing out everything on paper and storing the documents in a file cabinet. Others prefer to store information on disk so that it can be viewed on screen and printed out only as needed. Whichever method you choose, you should follow the following two important guidelines:

1. *Organize your information:* It's important to store documentation in an organized and logical manner. If the project is small, its documentation should be stored in a single folder (paper *or* electronic). Larger projects may require a group of folders, each relating to different aspects of the project. The folders should be labeled and dated with informative titles such as "The XYZ Project" and kept in a place that will be easy to find should another authorized person need to find it.

2. *Back up your information:* It's equally important to store a duplicate copy of all documentation. This guideline applies to written as well as electronic information. Fire, flood, theft, misplacement, and the all-too-common disk crash all can lead to the loss of a project's documentation trail. Archival storage of records in a different physical location will help to keep a project on track if one of these catastrophes should occur.

The Engineers's Logbook

One important vehicle for record keeping is the *engineer's logbook,* sometimes called the *engineer's notebook.* A well-maintained logbook serves as a permanent record that includes all ideas, calculations, innovations, and test results that emerge from the design process. When a project is brought to completion, all related logbooks are placed in an archive and remain the property of the company. An engineering notebook thus serves as an archival record of new ideas and engineering research achievements *whether or not they lead to commercial use.* A complete logbook serves as evidence of inventorship, establishes the date of conception and "reduction to practice" of a new idea, and shows that the inventor (you!) has used diligence in advancing the invention to completion. In this respect, the engineer's logbook is more than just a simple lab notebook. It serves as a valuable document that has legal implications. When you work as an engineer, you have a professional responsibility to your employer, your colleagues, and to the integrity of your job to keep a good logbook.

The notebook shown in Figure 3.11 is typical of many used in industry, government labs, and research institutions. It has permanently bound and numbered pages, a cardboard cover, and quadrille lines that form a coarse grid pattern. A label fixed to the front cover uniquely identifies the notebook and its contents. The company, laboratory, or project name is printed at the top, and the notebook is assigned a unique number by the user. In some companies and large research labs, a central office assigns notebook numbers to its employees when the notebook is signed out.

The techniques for logbook use are different from the procedures applied to simple lab notebooks, such as those found in your science classes in school. In many courses, instructors encourage students to write things down first on scratch paper and then to recopy relevant items from loose sheets into a neat notebook. This procedure is bad practice for an engineer. Although notebooks prepared in this way are easier for instructors to grade, the finished notebook seldom resembles a running record of what went on in the laboratory and is not especially useful for engineering design projects. Design is as much a *process* as it is a final product, and the act of writing things down as they happen helps you with your thinking and creativity. Also, keeping a record of what

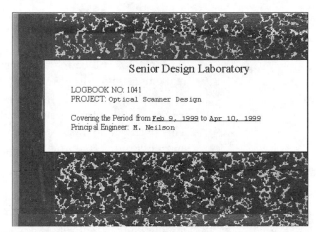

Figure 3.11.   Cover label of a typical engineers logbook.

did *not* work is just as important as recording what did, so that mistakes will not be repeated in the future.

Logbook Format

An engineering logbook should be used as an design *tool*. Enter everything into your logbook, no matter how seemingly irrelevant. Write down ideas as you think of them, even if you have no immediate plans to pursue them. Keep an ongoing record of successes and failures. Record the results of every mechanical, structural, electrical, system, flight, or performance test, even if the results are not used in the final design. Stopping to write things down will take discipline but will be worth the trouble at later stages in the design process. All-important information, including some you may have forgotten along the way, will have been entered into your logbook and will be at your fingertips to use as needed. Any format that meets your needs is suitable, as long as it forms a permanent record of the design process. Ban loose paper from the laboratory. It is easily lost, misplaced, or spilled upon. Resist the temptation to reach for the closest piece of loose paper when you need to do a calculation, record information, draw a sketch, or discuss an idea. Take the time to open your logbook, and use its pages for writing. You'll be glad you did when those numbers and sketches you need are readily available. Unbound paper used for anything other than doodling has no place in an engineering laboratory.

Using Your Engineer's Logbook

As the chief author of your logbook, you have the freedom to set your own objectives for its use. The following guidelines, however, are typical of those used by many engineers:

1. Each person working on a project should keep a separate logbook specifically for that project. All relevant data should be entered. When the logbook is full, it should be stored in a safe place specifically designated for logbook storage. In that way, everyone will know where to find the logbook when it's needed.

2. All ideas, calculations, experiments, tests, mechanical sketches, flow charts, circuit diagrams, etc. related the project should be entered into the logbook. Entries should be dated and written in ink. Pencil has a nasty habit of smudging or wearing away over time when pages rub against one another.

3. Logbook entries should outline the problem addressed, tests performed, calculations made, and so forth, but subjective conclusions about the success of

the tests (e.g., I believe . . .) should be avoided. The facts should speak for themselves. Logbook entries should not be a tape recording of your opinions.

4. The voice of the logbook should speak to a third-party reader. Assume that your logbook will be read by your boss, co-worker, or perhaps someone from marketing, who will review your work at some future date.

5. In company, corporate, or government settings the concluding page of each section or laboratory session should be dated and, where appropriate, signed. This practice eliminates all ambiguity with regard to the dates of invention and disclosure. Important entries should be periodically and routinely witnessed by at least one other person, preferably two. Witnesses should endorse and date the relevant pages with the words, "witnessed and understood."

6. Logbook pages should not be left blank. If a portion of a page must be left blank, a vertical or slanted line should be drawn through it. Pages should be numbered consecutively and not be torn out.

7. Relevant computer-generated plots, graphics, schematics, or photos printed on loose paper should be pasted or taped onto bound logbook pages. This procedure will help prevent loss of important data.

8. Do not make changes by using correction fluid. Cross out instead. This precaution will prevent you from creating obscure or questionable entries should your logbook be entered as legal evidence in patent or liability actions. Although this precaution probably won't be relevant to logbooks you keep for college design courses, it's a good idea to make the procedure a habit at an early point in your training.

Logbook Example

The following example illustrates proper use of an engineering logbook. Imagine that the pages to follow outline the design of your vehicle for the Peak Performance Design Competition introduced in Chapter 2. The first page shows a preliminary sketch of a basic concept for the vehicle based on a simple moving wedge. (See Figure 3.12.) The second page contains some calculations that estimate the battery drain as the vehicle

Figure 3.12. Logbook entry: Moving wedge concept for competition vehicle.

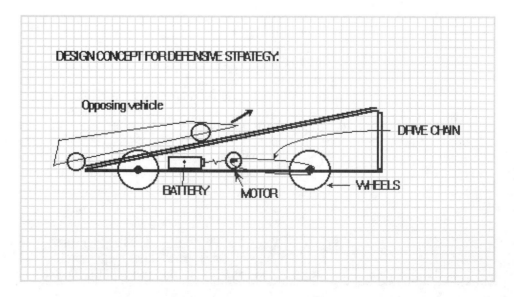

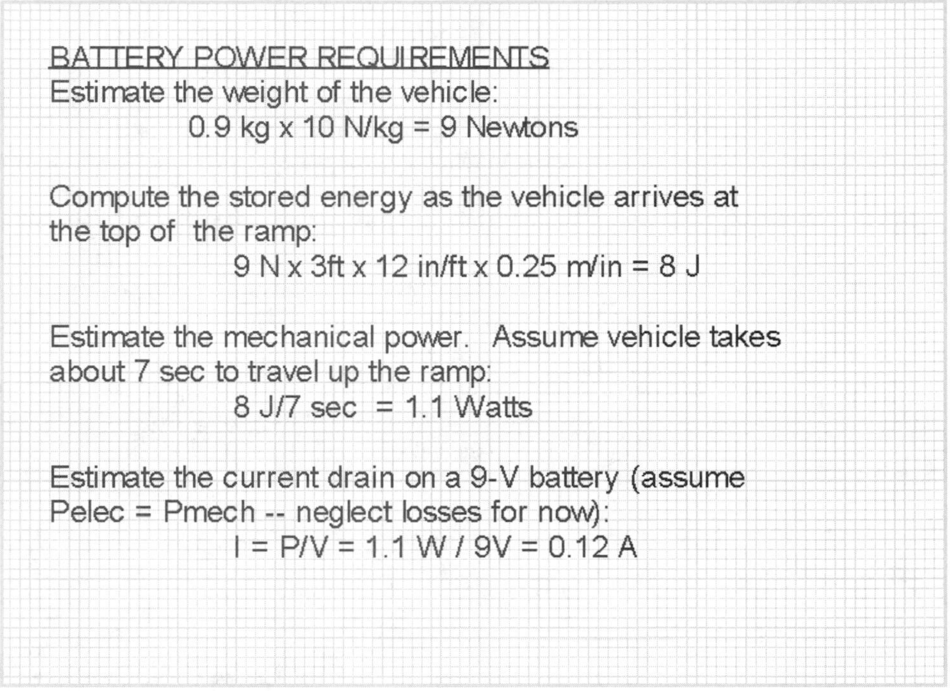

BATTERY POWER REQUIREMENTS
Estimate the weight of the vehicle:
        0.9 kg x 10 N/kg = 9 Newtons

Compute the stored energy as the vehicle arrives at
the top of the ramp:
        9 N x 3ft x 12 in/ft x 0.25 m/in = 8 J

Estimate the mechanical power.   Assume vehicle takes
about 7 sec to travel up the ramp:
        8 J/7 sec = 1.1 Watts

Estimate the current drain on a 9-V battery (assume
Pelec = Pmech -- neglect losses for now):
        I = P/V = 1.1 W / 9V = 0.12 A

Figure 3.13. Logbook entry: Power consumption calculations.

moves up the contest ramp. (See Figure 3.13.) The entries on the third page show a list of materials and parts to be purchased at the hardware store. (See Figure 3.14.) These parts will allow you to build a prototype and test your vehicle's ability to climb the ramp.

Dimensions and Tolerance

The sketch shown in Figure 3.15 shows the main chassis of the vehicle. Suppose that you were to have a machinist fabricate this part from a single, 0.4-cm-thick aluminum

Figure 3.14. Logbook entry: List of parts to be purchased at the hardware store.

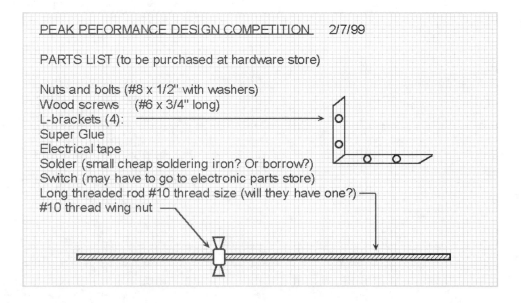

PEAK PEFORMANCE DESIGN COMPETITION   2/7/99

PARTS LIST (to be purchased at hardware store)

Nuts and bolts (#8 x 1/2" with washers)
Wood screws   (#6 x 3/4" long)
L-brackets (4):
Super Glue
Electrical tape
Solder (small cheap soldering iron? Or borrow?)
Switch (may have to go to electronic parts store)
Long threaded rod #10 thread size (will they have one?)
#10 thread wing nut

plate. Such a job requires specialized tools, including a *milling machine,* a *drill press,* and a *tap set.* Carefully note the labeled dimensions shown in Figure 3.15. These numbers communicate to the machinist the acceptable deviation, or *tolerance,* for each of the plate's various dimensions. No part can ever be made to *exact* dimensions, because machine tools do not cut perfectly. A cutting tool wanders about its intended position during the machining process. Similarly, changes in temperature, humidity, or vibration during the cutting process can cause the tool to follow a less-than-perfect path. The tolerance of each dimension shown in the drawing specifies the degree of error that will be acceptable for the finished product. As a rule, creating parts with tight tolerances involves the use of more expensive machining equipment and more time, because material cuts must be made more slowly. These features add considerable expense to the finished part. As the designer, you must decide which dimensions are truly critical.

The numbers on the drawing in Figure 3.15 have precise meaning for any machinist who reads the tolerance table. In this case, only the holes that will hold the axle mounts are especially critical. The length of the chassis base, for example, is 25 cm. The numbers 25.0, 25.00, and 25.000, though all mathematically equivalent, would mean

Figure 3.15.   Main chassis plate with dimensions and tolerance table.

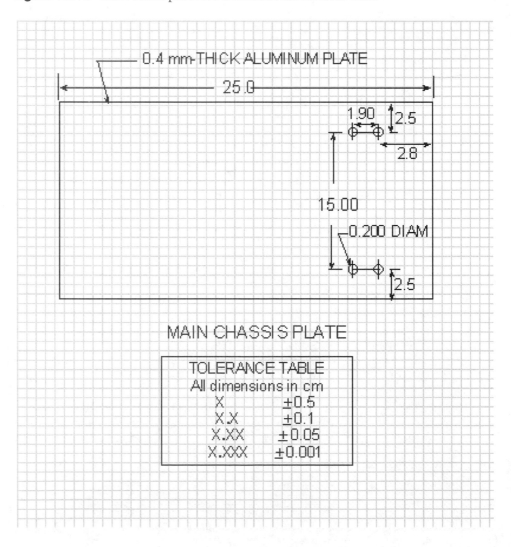

different things to the machinist. According to the tolerance table, the number 25.0, with one digit after the decimal point, should be interpreted by the machinist to mean 25 ± 0.1 cm. A chassis plate with a finished width diameter anywhere between 25.1 and 24.9 cm would be deemed acceptable. Similarly, the location of the holes that will hold the axle mounts are specified as lying 15.00 cm apart, implying a machined tolerance of 15 ± 0.05 cm. The minimum and maximum tolerance limits for the hole centers as machined would be between 15.05 cm and 14.95 cm. According to the tolerance table, the most stringent dimensions of all are those of the holes themselves. Because they will hold pins inserted by a friction fit, their diameters are specified to three decimal points, implying a strict machining tolerance of 0.200 ± 0.001 cm.

Significant Figures

The accuracy of any number used in technical calculations is specified by the number of *significant figures* that it contains. A significant figure is any nonzero digit or any leading zero that does not serve to locate the decimal point. A number cannot be interpreted as being any more accurate than its least significant digit, nor should a quantity be specified with more digits than justifiable by its measured accuracy. The numbers 128.1, 1.5, and 5.4, for example, imply numbers that have known accuracies of ±0.1, but the first is specified to four significant figures, while the second and third are specified to only two. If trailing zeros are placed *after* the decimal point, they carry the weight of significant figures. Thus the number 1.000 means 1 ± 0.001.

The accuracy of any computation can only be deemed as accurate as the *least* accurate number entering into the computation, and the number of significant figures that can be claimed for the result should be set accordingly. For example, the product 128.1 × 1.5 × 5.4 entered into a calculator produces the result 345.87. But because 1.5 and 5.4 are specified to only two significant figures, the rounded-off result of the multiplication must be recorded as 350, also with two significant figures. Note that a digit is rounded *up* if the digit to its right is 5 or more; if the digit to its right is less than 5, the digit is rounded down.

Technical Reports and Memoranda

Logbooks provide but one method of keeping a good documentation trail. Engineers also communicate information by writing technical reports at the significant milestones of a design project. A technical report describes a particular accomplishment, perhaps providing some project history or background material before explaining the details of what was achieved. The report may contain theory, data, test results, calculations, design parameters, or fabrication dimensions. Technical reports help form the backbone of a company's or laboratory's technical database and typically are stored in archival format, each with its own title and catalog number. Information for a technical report is gathered easily from a logbook this is accurate and up to date. When the time comes to write a journal paper, patent application, or product application note, the technical report becomes an indispensable reference tool. Techniques for writing technical reports in a clear and concise manner are presented in Chapter 7.

A technical report is also an appropriate way to explain why a particular idea did not work or was not attempted. Taking the time to write a technical report about a negative result or design failure can save considerable time later on should a design concept be revisited or attempted by engineers who were not present when the original project took place.

Schematics and Drawings

Documentation does not always appear in the form of text. Graphical records, such as drawings, circuit schematics, photographs, and plots, also become part of the documentation trail. These items are typically created with the help of computer software tools.

If paper is chosen as the storage medium, then graphical output should be printed on paper and kept in a folder along with other written records. If an electronic storage medium is preferred, then all files related to a particular project should be stored on disk in a logical hierarchy. Some engineers choose to keep all files for a project in a single file folder on the computer. Others prefer to sort files by the applications that produce them (e.g., CAD drawings in one folder, spreadsheet files in another, etc.) Yet other engineers like to transfer all the computer files related to a given project to a single removable disk that can be stored in a physical file-cabinet folder. Regardless of which storage method is chosen, the information related to a particular project should be carefully preserved in a format that will prevent loss.

Software Documentation

Of all design endeavors, the writing of software is one most prone to poor documentation. The revision loop of a software design cycle can be extremely rapid, because the typical software development tool allows the programmer to make small changes and test their effects immediately. This rapid-fire method of development invites poor documentation habits. Seldom does the software engineer find a good time to stop and document the flow of a program, because most pauses are short and change is frequent. As a result, the documentation for many software programs is added after the fact, if at all.

If you find yourself writing software, get into the habit of including documentation in your program as you go along. All software development tools provide a means for adding comment lines right inside the program code. Add them frequently to explain why you've taken a certain approach or written a particular section of program code. Explain the meaning of object names and program variables. Outline the flow of the program and the format of input data, output data, and graphical interfaces. Your in-program documentation should enable another engineer to completely understand and take over the writing of your program simply by reading the comment lines. Good in-program documentation will also be invaluable to you should you need to modify your program at a later time. It's amazing how quickly a programmer can forget the internal logic of a program after setting it aside for only a short time.

If your program is destined for commercial sale, then good internal documentation will easily translate into an instruction manual for the software. Better yet, write the instruction manual as you write the program code. You can change the instruction manual at the same time that you make major changes in the program code. The abundance of commercial software packages with pathetic or poorly written instruction manuals is testimony to generations of software engineers who have perpetuated a tendency toward poor documentation habits. If you master the skill of documenting software, your software products will be better utilized and more successful than those with poor documentation.

PROFESSIONAL SUCCESS: HOW TO KEEP GOOD RECORDS ALL THE TIME

If you want to keep a good documentation trail, get into the habit of carrying your logbook with you wherever you go. In that way, it will be available whenever you have a thought or idea related to your project. Buy a medium sized notebook that can fit easily into your backpack, if you carry one. Clip a pen right inside the front cover. Be sure to write your name and contact information on the cover in case you misplace your logbook! A tiny, $3 \times 5''$ bound notebook also will do nicely because it will readily fit into your pocket or purse. Although writing space will be limited because of the smaller size, you'll be more apt to have your logbook with you when an idea comes to mind.

Exercises 3.3

E10. Refer to the logbook calculations of Figure 3.13. Revise the estimate of the current drain on the battery if the vehicle weight is 1.7 kg.

E11. Refer to the logbook calculations of Figure 3.13. Convert all quantities to metric units and rewrite the logbook page.

E12. Refer to the logbook calculations of Figure 3.13. Revise the estimate of the current drain on the battery if the ramp height is 2 m and its length is 4 m.

E13. Refer to the tolerance table on the logbook page shown in Figure 3.15. Compute the difference between the maximum and minimum permissible physical values for dimensions specified by the following numbers: 10.0 cm, 7.55 cm, 1 cm, 2.375 cm, 0.005 cm.

3.4 ESTIMATION AS AN ENGINEERING DESIGN TOOL

Engineering design and estimation go hand in hand. When beginning any new design task, it's always a good idea to test for feasibility by doing rough calculations of important quantities and parameters. A paper-and-pencil analysis of a proposed strategy may eliminate fatal flaws before the actual construction process begins. The calculations need not be elaborate or precise. In the age of calculators and computers, students sometimes feel that an answer with lots of digits implies a better or more accurate answer. In many cases, however, "back of the envelope" calculations done by hand (recorded in your logbook, of course!) are all that are required to determine the soundness of a design strategy.

An Estimation Example

The following example, based on the Peak-Performance Design Competition introduced in Chapter 2, illustrates the usefulness of estimation as a design tool. Suppose that you and your teammate have decided to build a car that is battery operated and propelled by an electric motor. In an effort to conserve weight, you have decided to operate the vehicle from a single 9-V "transistor radio" battery, if possible. This design choice takes advantage of the fact that each run up the ramp is short in duration (around 15 seconds), and teams are allowed to change batteries between runs. Your strategy will be to heavily tax each battery to its maximum output and change it between each run up the ramp. In order to determine whether such a strategy is feasible, you must estimate the power to be delivered by the battery as the vehicle travels up the ramp. If this required power exceeds the amount available from a single battery, you will need to alter your design strategy and use two or more batteries.

Calculating the Power Required from the Battery

A simple calculation will reveal the power flow required from the battery as the vehicle travels up the ramp. The competition rules specify a maximum weight of 2 kg (about $4\frac{1}{2}$ pounds), but you hope to limit your vehicle weight to less than half that amount, or 0.9 kg (about 2 pounds). Of course, when you actually build your vehicle, you will need to determine whether your maximum weight target has been met.

Mechanical power for the vehicle will ultimately come from the battery. The motor has the job of converting electrical power into mechanical power. The electrical power going into the motor will have to equal the mechanical power transmitted to the wheels plus any electrical or mechanical losses in the drive train. This power–flow relationship is illustrated in Figure 3.16. Computing the mechanical power needed to

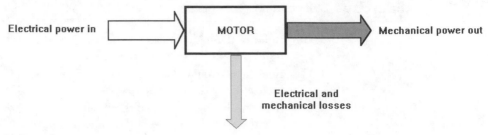

Figure 3.16.   Power flow diagram.

propel the vehicle up the ramp will allow you to estimate the electric power flow required from the battery. You first compute the gravitational force on the vehicle:

*Gravitational Force (Weight) Equals Mass
Times Gravitational Constant*

Newton's law states that

$$\mathbf{F} = m\mathbf{g},$$

where $\mathbf{g}$, the gravitational constant, is downward directed and has a magnitude of about 10 Newtons per kilogram. The weight of a 0.9-kg vehicle will be (0.9 kg)(10 N/kg) = 9 Newtons.

Energy Equals Force Times Distance

As the car is propelled to the top of the ramp, the mechanical energy supplied by the wheels will be equal to the net gain in the car's stored potential energy. This quantity will be equal to the magnitude of the gravitational force on the vehicle times the vertical distance traveled. The top of the hill lies 90 cm above the floor, hence the net gain in potential energy becomes

$$E = \|\mathbf{F}\|\Delta y = (9 \text{ N})(90 \text{ cm} \times 0.01 \text{ m/cm}) \approx 8 \text{ Joules.}^{*}$$

*Mechanical Power Equals Energy Per
Unit Time*

The quantity *power,* or flow of energy per unit time, is measured in joules per second. Assuming, for estimation purposes, that the run up the ramp lasts about 7 seconds, or half the allotted run time of 15 seconds, the mechanical power supplied by the wheels can be estimated by dividing the stored energy by the time required to climb the ramp:

$$P_{\text{mech}} = E/t = (8 \text{ J})/(7 \text{ s}) = 1.1 \text{ watts}$$

---

*The net gain in potential energy also can be expressed as the vector *dot product* of the vertically directed force and the trajectory *s* of the vehicle along the angle of the ramp. The ramp is inclined at an angle θ relative to the vertical, where θ can be computed from the ramp specifications. As shown in Figure 2.5, the vertical rise of the ramp is 90 cm, and the path length traveled by the vehicle is 30 cm + 120 cm = 150 cm. Hence,

$$\cos\theta = (\text{height})/(\text{path length}) = (90\text{cm})/(150 \text{ cm}) = 0.6$$

$$\Rightarrow \theta = \cos^{-1} 0.6 \approx 53°$$

The potential energy stored in the vehicle then becomes

$$E = \mathbf{F} \cdot \mathbf{s} = (9 \text{ N})(150 \text{ cm} \times 0.01 \text{ m/cm}) \times \cos 53° \approx 8 \text{ joules.}$$

This result is the same one computed from $\|\mathbf{F}\|\Delta y$.

A calculation such as this one should always be examined to make sure that the answer is reasonable. As a basis for comparison, consider a small plug-in night-light that draws about 4 watts. Expecting the car to draw about one quarter of that amount for 7 seconds indeed seems reasonable; the answer is believable.

Electrical Power Equals Current Times Voltage

The electrical power supplied to the motor must be equal to the mechanical power supplied to the wheels plus any losses that occur in the motor and drive train (e.g., the gears, belts, pulleys, bearings, and axle mounts). Neglecting these losses, one arrives at the simple conclusion that $P_{elec} = P_{mech}$. The electrical power supplied by the battery will be equal to the product of its voltage and its current, i.e., $P_{elec} = VI$. Hence, at a fixed voltage of 9 V, a power drain of 1.1 W will require a battery current of

$$I = P/V = (1.1 \text{ W})/(9 \text{ V}) \approx 120 \text{ mA}.$$

You next decide to obtain some information about batteries. Accessing the Duracell™ and Eveready™ Web pages* reveals that the typical 9-V battery can supply about 100 mA of current for short (1 hour or less) periods. This level of battery performance places your design specification at the border of feasibility. You can decide to use one battery only, possibly taxing it to its limit, or change your design specifications to include *two* batteries, each providing half the required power. For the moment, you decide that one battery is the better choice from the size and weight points of view. The 100 mA battery capacity is a general guideline, not an absolute limit. If you stick with your strategy of replacing the battery after every run, it might perform adequately. Also, if you increase the run time to 10 seconds, the power required will be reduced to 0.8 W, and the current required will be reduced to 90 mA.

On Second Thought

After reviewing your calculations and assumptions, you and your teammate realize that you have neglected all losses in the system. In reality, the conversion efficiency from electrical to mechanical power will be far from perfect. According to your professor, one might expect a power conversion efficiency of up to 90% from a well-designed, expensive motor, but you've decided to buy an inexpensive motor from a local electronic parts store to save money. Similarly, no more than about 60 percent of the converted mechanical power supplied by the motor shaft will show up at the wheels, because of frictional losses in the gears, drive belts, and axle mounts, leaving only about 50 percent of the power taken from the battery to actually propel the vehicle. The two-battery approach seems to be the more reasonable choice after taking losses into account.

Another Estimation Example (Not Related to the Peak-Performance Design Competition)

Imagine that you work for a company that makes automobiles. The head of manufacturing thinks that the company could save a lot of money by adopting a finishing process that impregnates color right into the body material during manufacturing using an electrostatic powder coating and baking technique. You are trying to convince your boss that the paint does not cost that much compared to the cost of making the rest of the car.

*http://www.duracell.com, http://www.eveready.com

The comparison is a tough call, because labor and equipment depreciation, not materials, are the main costs involved in most manufacturing processes. The overhead at a typical fabrication facility, including benefits, insurance, physical plant (the cost of keeping the factory open so that workers can do their jobs), and depreciation can run anywhere from 60 percent to 200 percent of salaries and other direct costs. Alternatively, the car could be painted by robot as are many large consumer items. In that case, labor costs would be eliminated, and a comparison between the powder and paint coating mechanisms is valid, because the cost of materials becomes the primary issue. The head of manufacturing has asked you, a manufacturing engineer, to estimate the total cost of conventional paint needed to cover the vehicle. How can you arrive at such an estimate? The steps required are outlined in the following discussion:

Draw a Rough Sketch of the Surfaces to Be Painted

The first step should be to draw a rough sketch of the car body on paper. One such sketch that depicts the various car surfaces is shown in Figure 3.17. The largest areas to be painted include the hood, trunk, roof, and both side fenders. The paint required to cover the posts of the window frames is negligible.

Estimate the Area of Each Section

Next estimate the area of each separate section of the car. The hood forms an approximate 1.2 m $\times$ 1.2 m square for a total area of about 1.4 square meters. The trunk is also nearly rectangular, measuring about 1.2 m $\times$ 1.5 m, for an additional 1.8 square meters. The doors are about 1 m long by 0.8 m tall, for a total area of 0.8 square meter each. The dimensions of the roof are about 1.4 m long by 1.2 m wide, for total of 1.7 square meters. For estimation purposes, each of the fenders can be modeled by one of the shapes shown in Figure 3.18. The surface area of the windows need not be counted, because they are not painted. The area of each fender can be calculated from the area formula for a trapezoid:

$$A = bc + c(a - b)/2.$$

If we assume reasonable numbers for the dimensions of the fenders, for example $a \approx 0.8$ m, $b \approx 0.4$ m, and $c \approx 1.5$ m for the front, and $a \approx 0.8$ m, $b \approx 0.4$ m, and $c \approx 1$ m for the rear, then we can compute the individual area estimates and add them to obtain an estimate for the total surface area of the car:

Figure 3.17.   Rough sketch of car body.

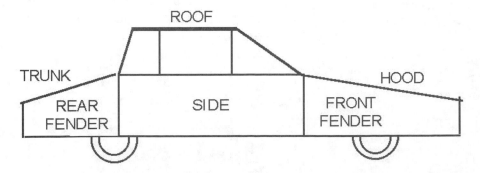

SIDE FENDER

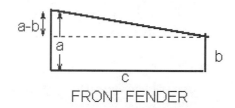

FRONT FENDER

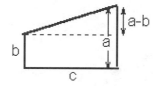

REAR FENDER

Figure 3.18.   Estimated shape of side fenders.

Hood: $1.2 \text{ m} \times 1.2 \text{ m} \approx 1.4 \text{ m}^2$

Trunk: $1.2 \text{ m} \times 1.5 \text{ m} \approx 1.8 \text{ m}^2$

Doors: $2 \times 1 \text{ m} \times 0.8 \text{ m} \approx 1.6 \text{ m}^2$

Roof: $1.4 \text{ m} \times 1.2 \text{ m} \approx 1.7 \text{ m}^2$

Front fenders: $2 \times [(0.4 \text{ m})(1.5 \text{ m}) + (1.5 \text{ m})(0.8 \text{ m} - 0.4 \text{ m})] = 2.4 \text{ m}^2$

Rear fenders: $2 \times [(0.4 \text{ m})(1 \text{ m}) + (1.5 \text{ m})(0.8 \text{ m} - 0.4 \text{ m})] = 2 \text{ m}^2$

Total area: $1.4 \text{ m}^2 + 1.8 \text{ m}^2 + 1.6 \text{ m}^2 + 1.7 \text{ m}^2 + 2.4 \text{ m}^2 + 2 \text{ m}^2 = 10.9 \text{ m}^2$

The result is equal to about 10 square meters. Round numbers are appropriate because only a rough estimate is required.

Multiply by the Thickness of the Paint

We next estimate the volume of paint required to cover the car by multiplying its surface area by the paint thickness. Paint thicknesses are usually measured in mils (1 mil = 0.001 inch $\equiv$ 0.025 mm = 25×10^{-6} m.) A very light coating of paint is typically about 0.5 to 1 mil thick, while a very heavy coating might be as thick as 7 or 8 mils. The choice of thickness for estimation purposes will depend on the paint application and whether we want the estimate to be on the high side or the low side of the actual value. Let's choose an average value of 4 mils. For this thickness, the volume of paint required to coat the car can easily be calculated:

Volume = area $\times$ thickness $\approx$ (10 m²) $\times$ (4 mils) $\times$ (25×10^{-6} m/mil) = 0.001 m³, or about one liter.

Exercises 3.4

E14. Determine the power required of the battery for the Peak Performance vehicle if the ramp is only two thirds as high as the value specified in the competition guidelines presented in Chapter 2.

E15. Determine the power required of the battery for the Peak Performance vehicle if a set of four 1.5-V AAA cells is used instead of a single 9-V battery.

E16. How much mechanical power can be derived from a 6-V electric motor that is 85% efficient if its maximum allowed current is 1 A?

E17. How much internal heat will be generated by an electric motor that is 60% efficient if it provides 20 kW of mechanical power to an external load?

E18. Estimate the amount of paint required to cover a single wooden pencil.

E19. Estimate the length of a 90 minute audio cassette tape.

E20. Estimate the number of platforms needed to build a scaffolding shell that encircles the Washington Monument.

E21. Estimate the volume of water contained within the supply pipes of a one-story, single family house.

3.5 PROTOTYPING AND BREADBOARDING: COMMON DESIGN TOOLS

Designing anything for the first time requires careful planning and forethought. As discussed in previous sections, the design process should begin with an exhaustive consideration of all possible alternatives and an effort to assess feasibility through preliminary estimations, sketches, and approximations. In some cases, a computer simulation of the design may be in order. (See Chapter 4.) After these initial phases of the design effort, it's time to proceed to the prototyping phase. A *prototype* is a mock-up of the finished product that embodies all its salient features but omits nonessential elements, such as a finished appearance or features not critical to the device's fundamental operation. Figure 3.19, for example, shows the prototype of a payload for a NASA rocket experiment.

If the product is an electronic circuit, it can be built up on a *breadboard*. A breadboard allows an engineer to wire together the various components of a circuit, such as resistors, capacitors, and integrated circuits, by plugging them into holes aligned with spring loaded clips located inside the breadboard. The clips make the electrical connections between components. In the prototype development stage, a breadboard readily permits changes and alterations to a circuit. A well-laid-out electronic breadboard is shown in Figure 3.20.

In production, the finished circuit is permanently soldered onto a printed wire board, such as those found inside computers, radios, and TVs. This form of circuit construction provides a durable, reliable finished product. An example of printed wire board construction is shown in Figure 3.21.

If the product is mechanical, its prototype should be fabricated from an easy-to-machine material that will permit testing, but perhaps not be as durable or visually attractive as the finished product. The chassis plate of Figure 3.15, for example, could be made from wood for initial tests of vehicle performance. Wood is easily drilled and formed, but is much less durable than metal for demanding applications. Mechanical prototypes can also be fabricated from various forms of angle iron and similar construction materials. One example is illustrated in Figure 3.22. The pieces of angle iron used to build this structure have holes in numerous places to allow for rapid construction and protyping.

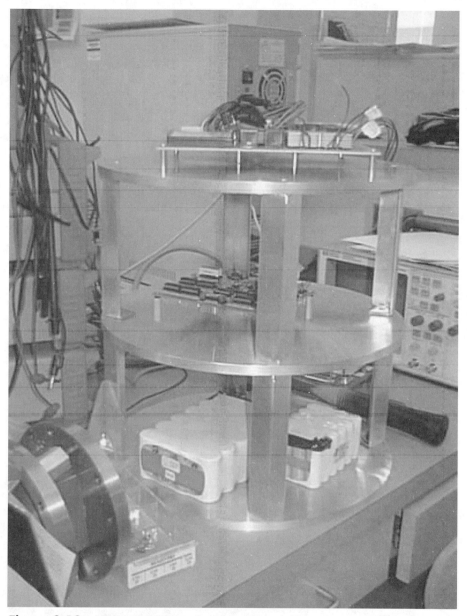

Figure 3.19. Prototype of the payload for a NASA rocket experiment.

Engineers and architects who design large structures, such as buildings, bridges, and dams, face a handicap not encountered by other engineers. It's virtually impossible to build a full-size prototype of these structures for testing purposes. It's not feasible, for example, to build the frame of a large sky scraper in some remote desert location to determine its maximum sustainable wind speed before collapse. Engineers who build large structures rely on *scale modeling* to guide them through the prototyping phase. Scale modeling relies on dimensional similarity to extrapolate observations made on a model of reduced size to the full-sized structure. Effects such as structural loading, wind loading, temperature, and large-scale motion are readily scaled and extrapolated. Wind tunnels are widely used to test scale models for aerodynamic effects. Vibration, combustion, and wave phenomena do not scale well because these effects are governed by

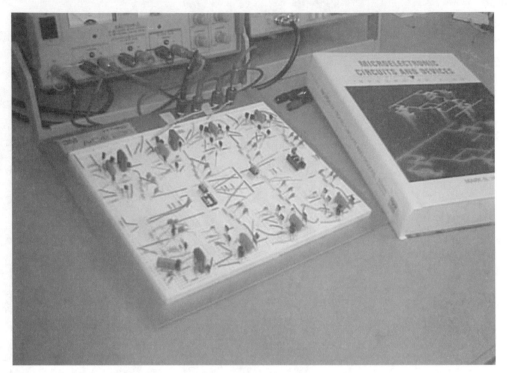

Figure 3.20. A well laid circuit on an electronic breadboard connected to two benchtop power supplies.

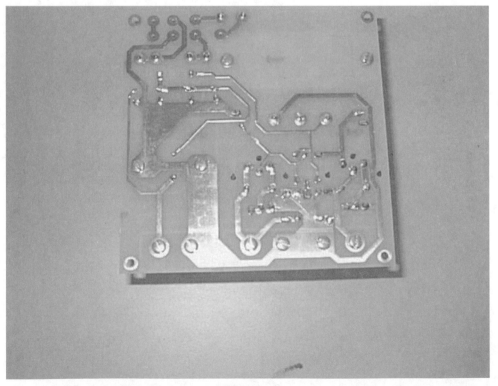

Figure 3.21. Underside view of a circuit fabricated on a printed wire board.

Figure 3.22. A prototype for testing an electrostatic chuck. The prototype has been fabricated from angle iron, plywood, and other simple materials.

physical parameters that are fixed regardless of the frame of reference. (That's one reason why movie scenes involving ships at sea or burning buildings often look unrealistic.)

Products designed for medical applications are developed in the prototype stage from inexpensive materials that exhibit the same physical characteristics as the finished product. A new design for a ball-and-socket replacement hip joint, for example, might be fabricated from aluminum for testing in a cyclical loading machine. The finished product ready for implantation would likely be produced from surgical-grade titanium at ten times the cost.

Software modules also undergo a prototype phase. Most software kernels (the part of a software program that does the actual "thinking") are surrounded by a graphical user interface (GUI) that provides access to the user. A software designer typically will write and test the kernel portions of a program long before writing the GUI.

A Prototyping Scenario

Prototyping is essential for the development and testing of electrical, mechanical, structural, and software devices. The prototype phase helps engineers reveal design flaws and problems that have escaped the initial planning and estimation phases. The following scenario illustrates the importance of careful prototyping in engineering design. It describes Tina's attempt to test her motor and timing circuit concept for the Peak-Performance Design Competition of Chapter 2. Tina's original ideas were outlined in the discussion on brainstorming presented in Section 3.2.

With the help of her teammate Juan, who prepared a prototype of the car's chassis frame from plywood, Tina mounted a small breadboard containing the car's electronic timing circuit and attached the drive motor to the chassis with some duct tape. She wired up the battery and connected it to the timing circuit. When she was finished with the wiring, Tina collected some diagnostic equipment: an oscilloscope to measure the output of the timing circuit, an ammeter to measure the motor current, and a voltmeter to monitor the voltage of the battery.

With the motor disconnected, Tina powered up the timing circuit and examined its output on the oscilloscope. Most of the circuit checked out, but Tina did find a wire that she had forgotten to connect between two of the integrated circuits. After adding the wire, and correcting one other crossed wire, the circuit appeared to function as she had intended. It produced a voltage for driving the motor that lasted for about 7 seconds, the precise time (according to her estimation) needed for the car to travel up the ramp.

Satisfied with the operation of the timing circuit, Tina next connected the motor to the load side of the circuit and turned it on. With the drive belt disconnected, the motor turned nicely and stopped after 7 seconds, so she and her teammate mounted the wheels, the gearbox, and the drive shaft and connected them to the motor with a small drive belt. The motor was about the size of a large D-cell battery. Tina held the car in midair. The motor again turned at a steady speed and cut out quickly when the timer circuit finished its timing pulse.

Tina next took the prototype and tried to run it up a large plank, which served as a mockup for the competition ramp. The plank, purchased at a lumber store, had one end propped up on a chair to simulate an inclined ramp. After turning on the switch, the car prototype traveled about half a car length and then stopped. At first, Tina was discouraged by this result. Was the timer faulty, or had she overlooked some fundamental design principle that would make her approach to the competition unworkable? She decided to perform further tests on the vehicle off the ramp. She clamped the car chassis in a vise with its wheels held in midair, connected the oscilloscope to the output of the timer, and connected the voltmeter across the battery and ammeter in series with the motor. When the switch was turned on, the motor turned as before for the entire 7-second duration of the timer signal. She noticed that the current to the motor was about 40 milliamperes and that the voltage on the 9-V battery dropped to about 8.2 V when the motor was running. Suspecting a battery drain problem, Tina next applied some friction to the wheels with her hand so simulate the load on the car traveling up the ramp. The current to the motor jumped to 160 mA, and the battery voltage dropped to 3.9 V. This drop in voltage caused the timer circuit, which required a power source of at least 5 V, to cease functioning. Tina had been overloading the battery! She redesigned the layout of the chassis to accommodate a set of six 1.5-V, AAA batteries connected in series instead of a single 9-V battery. This modification was performed easily on the wooden mockup of the chassis frame. At the expense of some added weight, the additional battery capacity could provide up to 200 mA of current with a voltage drop to only 8.2 V. Tina's battery performance problem was detected in the prototype phase.

PROFESSIONAL SUCCESS: WHERE TO FIND PROTOTYPING MATERIALS

The materials needed for creating prototypes can be found in many places. For building mechanical structures, no better place exists than your local hardware store. A well stocked hardware store sells all sorts of nuts, bolts, rods, dowels, fasteners, springs, hinges, and nails. Many common electrical parts, such as wires, terminal connectors, and switches, also may be found at the local hardware store. A home center—a large hardware store, lumber yard, plumbing supply, electrical supply, and garden shop all rolled into one—is an excellent source of structural prototyping materials such as plywood, angle iron, pipe, and brackets.

A selection of very basic electronic parts and breadboards can be found at hobby stores such as Radio Shack™, while comprehensive selections and competitive prices can be found in mail order and industrial catalogs. Examples include Digikey (www.digikey.com), Marlin P. Jones (www.marlin.com), Hosfelt (www.hosfelt.com), and Jameco Electronics (www.jameco.com).

3.6 REVERSE ENGINEERING

Reverse engineering refers to the process by which an engineer dissects a product to learn how it works. This design method is particularly useful if your goal is to duplicate a competitor's product or create a similar one using your own technology. Reverse engineering is practiced on a regular basis by companies worldwide. Although it may appear to be an unfair practice, in reality it can be a good way to avoid patent infringement and other legal problems by specifically avoiding an approach taken by a competitor. Reverse engineering one of your own products can be a good way to understand its operation if its documentation trail has been lost or is inadequate. Reverse engineering is encouraged in the writing of Web pages on the Internet. All of the major Web browsers provide a means to view and decipher the Hypertext Mark-up Language (HTML) code, or language used to encode the Web page, when it has been loaded onto your computer. This practice fosters the open information environment that has become the hallmark of the World Wide Web. Other forms of software, however, such as programs written in C, MATLAB, Mathematica, MathCad, or Fortran, can be particularly difficult to reverse engineer, particularly if the software has been poorly documented. The multitude of flow paths and logical junctions typical of such software programs can lead to confusion on the part of the reader and make it hard to understand how a program operates.

3.7 PROJECT MANAGEMENT

Even the simplest of design projects must be properly managed if they are to be successful. A systematic approach toward the completion of design goals is always preferable to a random, hit-or-miss approach. Project management is, like all the topics covered in this chapter, a design skill that must be learned and mastered. While the subject of project management could (and does) occupy the contents of entire books, we discuss here briefly three principal project management tools that can be used for design projects likely to be encountered by students or entry-level engineers.

Organizational Chart

When engineers gather to work as a team on a design project, some degree of hierarchy is necessary. It would be nice to approach all projects as a simple group of cooperative colleagues, but inevitably some team members will be burdened with more work than others unless everyone's responsibilities are clearly spelled out. One vehicle for specifying the management structure of a design project is the *organizational chart*. An organizational chart indicates who is responsible for each aspect of the design project, and it also describes the hierarchy of the team and those to whom the team members report. Figure 3.23 illustrates a simple organizational chart that might be used by students in the Peak Performance Design Competition. In this particular case, no student acts as team leader, but instead two independent teams work with each other and report to the

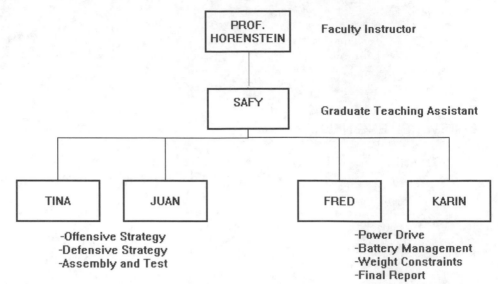

Figure 3.23.   Organizational chart showing responsibilities of the Peak Performance design team.

course teaching assistant for leadership and guidance. Other teams may choose to designate one student to act as overall team leader, while others may decide that several layers of administration are best. In the corporate world, where the structuring of workers and bosses can become complex, organizational charts are essential because each employee must understand to whom he or she reports and must know how upper management layers are structured.

Time Line

Time management is critical to the successful completion of a design project. In a perfect world, a design engineer would have as much time as necessary to work on all aspects of a design project, but in the real world, deadlines, demonstrations of progress, and the pressure to "get the product out the door" require that an engineer develop a sense of how much time will be needed for the various tasks of product development. (See, for example, the success box: The Laws of Time Estimation.) A *time line* can be a valuable tool for an engineer that wants to keep a project on schedule. A time line is simply a linear plot on which each of the various phases of a design project is assigned a milestone date. If any given task is in danger of not being completed before its designated milestone date, its the job of the engineer to allocate more time, and overtime if necessary, so that the task can be completed on time. A typical time line, such as one that a student might prepare for the Peak Performance Design Competition, is shown in Figure 3.24.

Gantt Chart

When a project becomes complex and involves many people, a simple time line may be inadequate for managing all aspects of the project. Similarly, if the project's various parts are interdependent, so that the completion of one phase depends on the success of others, the *Gantt Chart* of Figure 3.25 is a more appropriate time-management tool. The Gantt chart is simply a two-dimensional plot in which the horizontal axis is time, usually measured in blocks of days, weeks, or months, and the vertical axis represents either the tasks to be completed or the individuals responsible for those tasks.

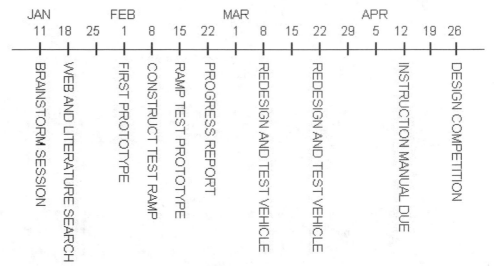

Figure 3.24.   Time line for scheduling tasks for the Peak Performance design team.

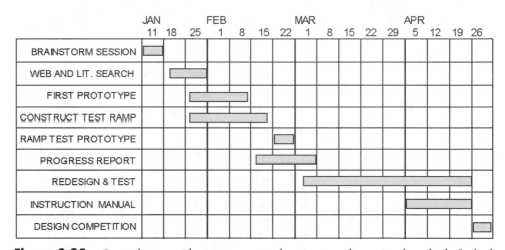

Figure 3.25.   Gantt chart provides a more comprehensive, two-dimensional method of scheduling the tasks for the Peak Performance design team.

Unlike the simple one-dimensional time line, which displays only the milestone dates for each phase of the project, the Gantt chart shows *how much* time is allotted to each task. It also shows the time overlap periods that are indicative of the interdependency between the various tasks of the project. When a particular task has been completed, it can be shaded in on the Gantt chart, so that the status of the project can be determined at a glance.

Exercises 3.7

E22.   Devise an organizational chart for building a float for the homecoming parade at your college or university.

E23.   Develop a time line for building a float for the homecoming parade at your college or university.

E24.   Develop a time line for a completing your course requirements over the time span of this semester.

E25.   Develop a Gantt chart that can help you plan to host an educational conference on engineering design. Consider all needed arrangements, including food, transportation, lodging, and meeting facilities.

PROFESSIONAL SUCCESS: PROJECT MANAGEMENT—THE LAWS OF TIME ESTIMATION

How long will it take to perform a particular design task? How much time should I allot to each segment of the design cycle? The following three Laws of Time Estimation will help you to determine the time required for elements of the design process:

1.   Everything takes longer than expected
2.   If you've worked on something similar before, estimate the amount of time required to finish the task. The actual amount of time required will be about four times longer.

3.   If you've *never* worked on something similar before, estimate the amount of time required to finish the task. The actual amount of time required will be equal to the next highest time unit. (For example, something estimated to take an hour will take a day; something estimated to take a day will take a week, etc.)

KEY TERMS

Brainstorming	Documentation	Estimation
Prototyping	Project management	

Problems

Brainstorming: Use brainstorming methods to generate solutions to the following problems:

1.   You are given a barometer, a stop watch, and a tape measure. In how many different ways can you determine the height of the World Trade Center in New York City?
2.   Design a sensing mechanism that can measure the speed of a bicycle.
3.   Many international airline flights allow smoking in the rear seats of the aircraft. Design a system that will remove or deflect smoke from the front seats of the aircraft.
4.   You are given an egg, some tape, and several drinking straws. Using only these materials, design a system that will prevent the egg from being broken when dropped from a height of six feet (two meters).
5.   Devise as many different methods as you can for using your desktop computer to tell time.
6.   Design a system for washing the inside surfaces of a large aquarium (the kind the public visits to see large sea creatures) from the outside.
7.   Design a system to be used by a quadriplegic to turn the pages of a book.
8.   Devise a system for automatically raising and lowering the flag at dawn and dusk each day.
9.   Design a system that will automatically turn on a car's windshield wipers when needed.
10.   Develop a device that can alert a blind person to the fact that water in a pot has boiled.
11.   Devise a system for lining up screws on an assembly-line conveyor belt so that they are all pointing in the same direction.
12.   Develop a method for detecting leaks in surgical gloves during the manufacturing process.
13.   Devise a method for deriving an electrical signal from a magnetic compass so that it can be interfaced with a computer running navigational software.
14.   Given a coil of rope and eight poles, devise a method to build a temporary emergency shelter in the woods.

15. Devise an alarm system to prevent an office thief from stealing the memory chips from inside a personal computer.

16. Imagine custodial workers who are in the habit of yanking on the electric cords of vacuum cleaners to unplug them from the wall. Devise a system or device to prevent damage to the plugs on the ends of the cords.

17. Develop a system for automatically dispensing medication to an elderly person who has difficulty keeping track of schedules.

18. Develop a system for reminding a business executive about meetings and appointments. The executive is always on the go, but can carry a variety of portable devices and gadgets. Feel free to use your knowledge of existing communications systems and technology, if necessary.

19. Devise a system that will agitate and circulate the water in an outdoor swimming pool so that a chlorine additive will be evenly distributed. Assume that an electrical outlet is available at the site of the pool.

20. Devise a system that will allow a truck driver to check tire air pressure without getting out of the vehicle.

Logbook and Record Keeping:

21. Begin to keep a logbook for your class activities. Enter sketches and records of design assignments, inventions, and ideas.

22. Pretend that you are Alexander Graham Bell, the inventer of the telephone. Prepare several logbook pages that describe your invention.

23. Pretend that you are Marie Curie, the discoverer of the radioactive element radium. Prepare several logbook pages that describe the activities leading to your discovery.

24. Pretend that you are Dr. Zephram Cockran, the inventor of plasma warp drive on the television and movie series Star Trek®. Prepare several logbook pages that describe your invention.

25. Imagine that you are Elias Howe, the first inventor to perfect the sewing machine by putting the eye of the needle in its tip. This innovation made possible the bobbin system still in use today. Prepare several logbook pages that describe your invention and its initial tests.

26. Reconstruct logbook pages as they might have appeared for the person inventing the common paper clip.

27. Evaluate each of the following numerical computations, expressing the result with an appropriate number of significant figures:

    a. $F = 1221 \text{ kg} \times 0.098 \text{ m/s}^2$

    b. $V = 56 \text{ A} \times 1200 \text{ ohms } (\Omega)$

    c. $x = 76.8 \text{ m/s} \times 1.000 \text{ s}$

    d. $m = 56.1 \text{ lb} + 45 \text{ lb} + 98.2 \text{ lb}$

    e. $i = 91.4 \text{ V} \div 1.0 \text{ k}\Omega$

    f. $P = (5.1 \text{ V})^2/(1.0 \text{ k}\Omega)$

28. When calculations are performed, the answer will only be as accurate as the weakest link in the chain. An answer should be expressed with the same number of significant figures as the least accurate factor in the computation. Express the result of each of the following computations with an appropriate number of significant figures:

    a. $V = (12.9 \text{ mA})(1500 \text{ }\Omega)$

    b. $F = 2.69 \text{ kg} \times 9.8 \text{ m/s}^2$

    c. $F = -3.41 \text{ N/mm} \times 6.34 \text{ mm}$

    d. $i_B = (1.29 \text{ mA})/(100)$

    e. $Q = (6.891 \times 10^{-12} \text{ F})(2.34 \times 10^3 \text{ V})$

29. Measure the dimensions of an ordinary 3.5" computer diskette. Prepare a dimensioned sketch of the diskette, complete with tolerance table.

Estimation:

The following three problems relate to the first estimation example on page 73 concerning the amount of power required to propel the vehicle up the ramp in the Peak-Performance Design Competition:

30. For the chosen approximate run time of 7 seconds, will the battery *always* supply 1.1 W of power to the vehicle? What would be a reasonable estimate of the average power flow over the 15-second run time?

31. Suppose that smaller batteries are chosen that can supply only 50 mA of peak current. Such a decision might be made to reduce battery weight and produce a lighter vehicle. If vehicle weight is reduced to 0.5 kg, what power flow can be expected? What will be the peak battery current?

32. If motors are chosen with 95 percent efficiency, and mechanical losses are 60 percent, what will be the required battery current?

The following four problems involve vector addition. Vector manipulation is an important skill for estimating forces in mechanical systems. When adding forces or other quantities represented as vectors, the principles of vector addition must be followed. Vectors to be added are first decomposed into their respective x-, y-, and z-components. These components are added together separately, then recombined to form the total resultant vector. Sometimes it's convenient to decompose vectors into components lying on axes other than the x-, y-, and z-axes.

33. Two guy wires securing a radio antenna are connected to an eye bolt. One exerts a force of magnitude 3000 N at an angle of 10° to the vertical. The other exerts a force of 2000 N at an angle of 75° to the vertical. Find the magnitude and direction of the total force acting on the eye bolt.

34. A guy wire exerts a force on an eye bolt that is screwed into a wooden roof angled at 30° to the horizontal. The guy wire is inclined at 40° to the horizontal. If the eye bolt is rated at a maximum force of 1000 N perpendicular to the roof, how much tension can safely be applied to the guy wire?

35. A large helium-filled caricature balloon featured in a local parade is steadied by two ropes tied to its midpoint. One rope extending on one side of the balloon is inclined at 20° to the vertical. A second rope located on the other side of the balloon is inclined at 30° to the vertical. If the balloon has a buoyancy of 200 kN, what will be the tension in each of the ropes?

36. An eye bolt is fixed to a roof that is inclined at 45° relative to the x-axis. The eye bolt holds three guy wires inclined at 45°, 150°, and 195°, respectively, measured clockwise from the x-axis. These wires carry forces of 300 N, 400 N, and 225 N, respectively. What is the magnitude and direction of the total resultant force? What are the components of force measured perpendicular and parallel to the roof line?

Problems 37–54 can help you develop your design estimation skills. Discuss them with your friends, and see if you arrive at the same approximate answers.

37. Estimate the amount of paint required to paint a Boeing 747™ airplane.

38. Estimate the cost of allowing a gasoline-powered car to idle for 10 minutes.

39. Estimate the daily consumption of electrical energy by your dormitory, residence, apartment building, or home. (Check your estimate against real electric bills if any are available.)

40. Estimate the cost of leaving your computer running 24 hours per day.

41. Estimate the cost savings of installing storm windows on an average-sized four-unit apartment building.

42. Estimate the gross weight of a fully loaded eighteen-wheel tractor trailer.

43. Estimate the number of single-family houses in your home state.

44. Estimate the number of bolts required to assemble the Golden Gate Bridge.

45. Estimate the number of bricks in an average-sized house chimney.

46. Compute the surface area of all the windows in your dorm, apartment building, or house where you live.

47. Estimate the amount of carpet that it would take to cover the playing field at San Francisco's 3Com Park.

48. Estimate the total mass of air that passes through your lungs each day.

49. Estimate the time required for a stone to fall from sea level to the bottom of the lowest point in the Earth's oceans.

50. Estimate the cost of running a medium-sized refrigerator for a year.

51. Estimate the weight of a layer of shingles needed to cover a single-family, ranch-style house that has a flat roof. Revise your calculations for a pitched roof.

52. Estimate the physical length of a standard 120-minute VHS video cassette tape.

53. Estimate the number of microscopic pits on an average-sized audio compact disk (CD) or digital video disk (DVD).

54. Estimate the number of books checked out of your school library each week.

Prototyping:

55. Suppose that 100 mA of steady current flows from a 9-V battery via a timer circuit to a motor. If the controller circuit is 92 percent efficient and the motor 95 percent efficient, how much mechanical power is transferred to the motor wheels (neglecting bearing friction)?

56. Ohm's law states that the voltage across a resistor is equal to the current flowing through it times the resistor value $(V = IR.)$ Calculate the current flowing through each of the following resistors if each has a measured voltage of 24 V across it: 1 Ω, 330 Ω, 1 kΩ, 560 kΩ, 1.2 MΩ (Note: 1 kΩ $= 10^3$ Ω; 1 MΩ $= 10^6$ Ω).

57. Ohm's law states that the current flowing through a resistor is equal to the voltage across it divided by the resistor value $(I = V/R.)$ Calculate the voltage across each of the following resistors if each has a measured current of 10 mA: 1.2 kΩ, 4.7 kΩ, 9.1 kΩ, 560 kΩ, 1.2 MΩ. (Note: 1 mA $= 10^{-3}$ A, 1 kΩ $= 10^3$ Ω, and 1 MΩ $= 10^6$ Ω).

58. Kirchhoff's current law states that the algebraic sum of currents flowing into a common connection, or *node,* must sum to zero. Suppose that currents of 1.2 A, –5.4 A, and 3.0 A flow on wires that enter a four-wire node. What current must flow *out* of the fourth node?

59. Kirchhoff's voltage law states that the sum of voltages around a closed path must sum to zero. Three resistors are connected in series to a 9-V battery. The measured voltages across two of the resistors are 5 V and 2.5 V, respectively.

    a. What is the voltage across the third resistor?

    b. The first two resistors have values of 100 Ω and 50 Ω, respectively. What is the current flowing through the third resistor?

60. High-power devices, such as thyristors and power transistors, are often mounted on metal *heat sinks.* A heat sink enhances the overall thermal contact between the device package and the surrounding air, leading to a cooler device and a larger power-dissipation capability. Heat removal is important, because excess heat can cause a catastrophic rise in device temperature and permanent failure. Every heat sink has a heat-transfer coefficient, or *thermal resistance,* Θ (capital Greek theta), which describes the flow of heat from the hotter sink to the cooler ambient air. The ambient air is assumed to remain at constant temperature. This thermal flow can be described by the equation $P_{therm} = (T_{sink} - T_{air})/Θ$, where P_{therm} is the thermal power flow out of the device, T_{sink} is the temperature of the heat sink, and T_{air} the temperature of the air.

    a. A power device is mounted on a heat sink for which Θ $= 4.5$ °C/W. A total of 10 W is dissipated in the device. What is the device temperature if the ambient air temperature is 25°C?

    b. A device rated at 200°C maximum operating temperature is mounted on a heat sink. If the ambient air is 25°C and 25 W of power must be dissipated in the device, what is the largest thermal coefficient Θ that the heat sink can have?

61. A switch is a mechanical device that allows the user to convert its two electrical terminals from an open circuit (no connection) to a short circuit (perfect connection) by moving a

lever or sliding arm. A *switch pole* refers to a set of contacts that can be closed or opened by the mechanical action of the switch. A *single-pole, double-throw* (SPDT) switch has three terminals: a center terminal that functions as the common lead, and two outer terminals that are alternately connected to the center terminal as the position of the switch lever is changed. When one of the outer terminals is connected to the center terminal, the remaining outer terminal is disconnected from the center terminal.

a. Consider the problem of wiring the light in the stairway of a two-story house. Ideally, the occupants should be able to turn the light on or off using one of two switches. One switch is located at the top of the stairs, and the other is located at the bottom. Toggling either switch lever should make the light change state. Draw the diagram of a circuit that illustrates the stairway lighting system.

b. Now consider the problem of a *three*-story house in which the lights in the stairwell are to be turned on or off by moving the lever of any one of three switches (one located on each floor). Design an appropriate switching network using two single-pole switches and one double-pole switch. (A double-pole switch has six terminals and can be thought of as two single-pole switches in tandem, with both levers engaged simultaneously.)

62. A dc motor consists of a multipole electromagnet coil, called the *armature,* or sometimes the *rotor,* that spins inside a constant magnetic field called the *stator* field. In the small dc motors typically found in model electric cars and toys, permanent magnets are used to create the stator magnetic field. In larger, industrial-type motors, such as an automobile starter or windshield-wiper motors, the stator field is produced by a second coil winding.

   Current is sent through the rotating armature coil by way of a set of contact pads and stationary brushes called the *commutator.* Each set of commutator pads on the rotor connects to a different portion of the armature coil winding. As the rotor rotates, brush contact is made to different pairs of commutator pads so that the portion of the armature coil receiving current from the brushes is constantly changed. In this way, the magnetic field produced by the rotating armature coil remains stationary and is always at right angles to the stationary stator field. The north and south poles of these fields constantly seek each other, and because they are always kept at right angles by the action of the commutator, the armature experiences a perpetual torque (rotational force). The strength of the force is proportional to the value of armature current, hence the speed of the motor under constant mechanical load is also proportional to armature current.

a. Obtain a small dc motor from a hobby or electronic parts store. Connect two D-cell batteries in series with the motor without regard to polarity. Observe the direction of rotation, and then reverse the polarity of the battery connections. Observe the results.

b. As an engineer, you are likely to encounter situations in which the rotational direction of a dc motor must be changed by a switch control. Using a double-pole, double-throw switch like the one described in the previous problem, design a circuit that can reverse the direction of the motor using a single switch.

Reverse Engineering:

63. Take apart a retractable ball point pen (the kind that has a push button on top to extend and retract the writing tip). Draw a sketch of its internal mechanism, and write a short description of how the pen works.

64. Take apart a common 3.5" floppy computer diskette. Make a sketch of inner construction, and write a short summary of its various components.

65. Take apart a standard desktop telephone. Use your investigative methods to develop a block diagram of how the phone works and connects to the outside world.

66. Suppose that you have been given the assignment to design a desktop stapler. Dissect an existing model from a competitor, draw a sketch of its mechanical construction, and create a parts list for the stapler.

67. Take apart a common flashlight, draw a sketch of its mechanical construction, and create a parts list from which you could reproduce another.

68. Imagine that you have discovered an errant, unoccupied space vehicle in an outlying field. Write a report in which you reverse engineer the spacecraft to discover elements of its technology. Examine the vehicle's propulsion system, telemetry, and sensor systems.

Project Management:

69. Suppose that you've been given the assignment to write a research paper on the history of human air flight. Develop a time line for completing this comprehensive research assignment.

70. Create a Gantt chart for your own hypothetical entry into the Peak-Performance Design Competition.

71. Imagine that you work for a company that is designing an electric car for commercial sale. Create an organizational chart for the company and a Gantt chart for designing the vehicle's drive train.

72. Choose an engineering company with which you are familiar or in which you have an interest. Develop an organizational chart for the company. Information about a company's structure and personnel often can be found on the company's Web page.

73. Imagine that you wish to start your own company to write software tools for doing business on the Web. Create an organizational chart that outlines the positions you'll need to fill to get the company started.

4

The Computer as Part of the Design Solution

Computers have become so commonplace that it's hard to imagine life without them. They affect all aspects of technology, communication, business, commerce, government, education, finance, medicine, avionics, and social service. Computers have even become a form of recreation. One can debate the value of computers to human relationships and society at large, but in the world of engineering, computers have become indispensable. As do people in all professions, engineers use computers for communication, information processing, word processing, electronic mail, and Web browsing. But the true worth of the computer to the engineer lies in its ability to perform calculations extremely rapidly. A computer can be programmed with user-written code to perform all sorts of numerical calculations. Commercial simulation, spreadsheet, and graphing programs allow engineers to determine everything from the stresses on mechanical parts and the operation of complicated electronic circuits to the force loads on building frames and theoretical predictions of performance tests. The availability of the computer and the abundance of software tools that have proliferated over the past decade have greatly enhanced the productivity and capabilities of engineers in all disciplines. In this chapter, we examine some of the many uses of the computer in engineering design.

SECTIONS

- 4.1 When Are Computers Really Necessary?
- 4.2 The Computer as an Analysis Tool
- 4.3 Spreadsheets and Data Bases
- 4.4 Graphical Programming
- 4.5 The World Wide Web
- 4.6 Real-Time Computer Control
- 4.7 Analog-to-Digital and Digital-to-Analog Conversion

OBJECTIVES

In this chapter, you will:

- Examine the role of the computer in engineering design.
- Learn when and when not to use the computer.
- Discuss several examples of computer use for analysis, data collection, and real-time control.

4.1 WHEN ARE COMPUTERS REALLY NECESSARY?

Computers play an important role in solving engineering problems. Their use in virtually all technical disciplines has come to be expected, and the use of computers has become a mandatory part of most educational programs in engineering. Despite the highly beneficial symbiosis between design and the computer, a danger exists when use of the computer supersedes human creativity and judgment in the design process. This phenomenon is sometimes called the *can-do* trap. All too often, we spend time doing things on a computer simply because we *can* do them. We simulate mechanical structures or circuits and blindly accept the results without verifying the fundamental principles behind the simulations or comparing their output with real physical tests. We create documents that lack real substance, but are visually perfect composites of embedded graphics, color printing, and fancy fonts. We create laptop slide screens for presentations with moving arrows, blinking screens, sound, and video, but do an inferior job of verbally conveying our message when compared to a well-done presentation given with simple hand-drawn overhead slides. A good engineer learns to harness the power of the computer without falling into the can-do trap. The computer is a precision tool that should be used like a fine measuring instrument, and not as a sledge hammer. In engineering design, computers should become an adjunct to our own creative process, not our preoccupation. The examples of this chapter illustrate several ways in which the computer rightfully can be used as a meaningful part of the design process.

Exercises 4.1

E1. Discuss the way in which a computer might be used to automatically produce time lines for new design projects.

E2. Discuss the way in which a microprocessor might be used to build a digital alarm clock.

E3. Identify three appliances or machines that utilize the power of a microprocessor.

4.2 THE COMPUTER AS AN ANALYSIS TOOL

As discussed in Chapter 3, when engineers design something for the first time, they often build simplified prototypes to test the basic operating principles of the device. In many cases, simple hand calculations are all that are needed as a prelude to building a working prototype. At other times, however, more complex calculations are required to verify the feasibility of a design feature or concept. Computers can become an important part of this verification process. Simulation is one major task that computers perform extremely well, and numerous software programs exist that help engineers simulate everything from bridges to electronic circuits. Examples of popular simulation programs include PSPICE (electrical and electronic circuits), PowerView® (digital logic circuits), MATLAB®, Mathcad®, Mathematica® (general mathematics and matrix manipulation), Ansoft® (structural and field analysis), and Supreme® (semiconductor devices). The example in the following section illustrates the use of computers to perform engineering analysis. The analysis steps are first presented from a general point of view, and then illustrated using the MATLAB® programming language. Despite this choice of software environment, the methodology is universal and can be used with any computer language or software program capable of performing numerical calculations. Other math packages, such as Mathcad® and Mathematica®, or the programming language C, for example, could be used with equal ease. Subsequent sections of the chapter discuss spreadsheets, graphical user interfaces, and the use of computers in embedded and real-time control.

EXAMPLE 4.1:
TRAJECTORY
ANALYSIS

The Peak-Performance Design Competition was introduced in Chapter 2 and further developed in Chapter 3. In Section 3.2, Tina conceived of an offensive strategy based on a flying harpoon. (See page 59.) Imagine that Tina and Juan have decided to build the prototype of such a device. They plan to build the prototype using a launch tube, retractable rubber band, and harpoon that has a notched trailing end, as illustrated in Figure 4.1. One approach would be to first put together a crude prototype that "seemed about right" from whatever rubber bands, rods, and tubes might be lying around. The students could then test the prototype to see if it worked and fiddle with it to optimize its performance. This method certainly would draw upon their engineering intuition, as well as any previous practical experience they may have had in constructing launched projectiles. Indeed, it probably would be fun to march ahead and try to build a working harpoon launcher from the start. But a better approach, and one more likely to meet with ultimate success, would be to simulate first the operation of the launching mechanism on a computer to test its feasibility. Calculations of this sort would help determine the optimum values of the system's various design parameters before construction begins. In this case, the variables to be set include the weight, diameter, and length of the harpoon, the force constant and retraction distance of the rubber band used to launch the harpoon, and the angle of inclination of the launch tube. Other constraints include the dimensions of the ramp (fixed by the rules) and the maximum allowed size of the vehicle (30 cm × 30 cm × 30 cm.) These last dimensions determine the maximum length of the harpoon, which must fit within the allowed boundaries of the vehicle.

One possible trajectory for the harpoon is illustrated in Figure 4.2. Juan recalls from his physics classes that if aerodynamic forces are negligible, the shape of the harpoon's trajectory will be parabolic.° The choice for the harpoon's landing point, however, is somewhat arbitrary and becomes part of the students' offensive strategy. If they

Figure 4.1. Launch tube, stretched rubber band, and harpoon for Peak Performance offensive strategy.

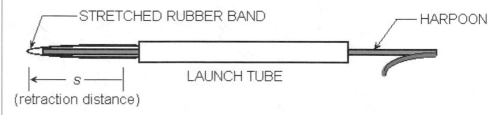

Figure 4.2. Desired parabolic trajectory and landing site for the harpoon.

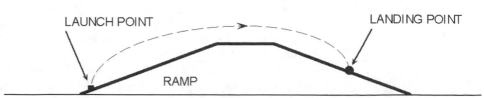

°When an object is propelled by an impulse force, it will acquire nearly instantaneously a velocity at $t = 0$. The horizontal component of velocity will remain fixed, while the vertical component will be subject to an acceleration due to gravity. This effect leads to a parabolic trajectory.

aim for a landing spot that lies past, but too close to the top of the hill, they may allow their opponent to push the harpoon the short distance needed to reach the top of the hill. If they locate the landing spot over but too far down the hill, they may allow their opponent to travel past the landing spot before the harpoon can be launched. Also, the farther away the landing spot is from the launch point, the greater the chance that small lateral (sideways) errors in launch angle will be amplified downstream along the harpoon's trajectory. These lateral errors might cause the harpoon to go over the edge of the ramp and land on the floor.

A computer can easily help Tina and Juan solve for the harpoon's trajectory. Variables of relevance include the harpoon's weight, the initial energy stored in the rubber band, and the angle of inclination of the launching tube. Before writing a computer program, however, they realize that they must examine the basic physics governing the harpoon's trajectory so that they can correctly code the program. Assuming that the rubber band, when released, will impart an initial velocity v_0 to the harpoon, the harpoon's equation of motion after ejection will be given by Newton's law of motion

$$\mathbf{F} = m\mathbf{a}$$

where $\mathbf{F}$ is the force of gravity, and $\mathbf{a}$ is the harpoon's acceleration. In this case, $\mathbf{F}$ acts in the y-direction only, hence, to the extent that aerodynamic forces can be ignored the x- and y-components of Newton's law can be integrated to yield

$$x = x_0 + v_{x0}t \tag{4.1}$$

and

$$y = y_0 + v_{y0}t - mgt^2/2 \tag{4.2}$$

Here x_0 and y_0 define the position of the harpoon (measured at its center of gravity) at $t = 0$. The variable m is the mass of the harpoon, and g is the gravitational constant. The quantity mg is the magnitude of the force due to gravity. Note that gravity does not affect the value of x. The harpoon's horizontal velocity is constant in time.

The harpoon's exit velocity at $t = 0$ will be determined by the potential energy stored in the stretched rubber band. The latter can be computed from the band's force equation. For a rubber band exhibiting a linear restoring force, the force equation will be given by

$$F_{band} = -ks \tag{4.3}$$

Here F_{band} is the force produced by the stretched band, k is the band's spring constant in newtons per meter (N/m), and the stretch s is the length that the band has been elongated from its unstretched position. The potential energy E_P stored in the stretched band can be found by integrating the force equation as a function of s, yielding

$$E_P = -\int F \, ds = \int ks \, ds = ks^2/2 \tag{4.4}$$

When the harpoon exits the launch tube, all of the potential energy stored in the rubber band will be converted to kinetic energy of the moving harpoon (assuming that frictional losses in the launch mechanism are negligible). The kinetic energy of the harpoon can be expressed as

$$E_K = mv_0^2/2 \tag{4.5}$$

where v_0 is the harpoon's exit velocity. Equating E_P to E_K results in an expression for v_0 as a function of band retraction s:

$$v_0 = \sqrt{k/m} \; s \tag{4.6}$$

The horizontal and vertical components of this exit velocity can be expressed by

$$v_{x0} = \sqrt{k/m}\ s\ \cos\theta \tag{4.7}$$

and

$$v_{y0} = \sqrt{k/m}\ s\ \sin\theta, \tag{4.8}$$

where θ is the angle of inclination of the launch tube. Using Eqs. (4.1), (4.2), (4.7), and (4.8), one can calculate the evolution of the trajectory x,y of the harpoon as a function of time. In principle, Tina and Juan could solve these equations in closed algebraic form to find the exact landing spot on the ramp. Such a calculation would be very tedious to perform. An alternative method is simply to use the computer to plot the harpoon's trajectory and to observe where it lands on the ramp. Although this second method will not provide as accurate an answer as the first one, it has visual appeal and is well suited for implementation on a computer. A generic flowchart illustrating the calculation method is shown in Figure 4.3.

Before any simulation can be performed, Tina and Juan must determine the various fixed parameters of the harpoon and the launching system. They've selected a weight of 100 gm (0.1 kg) for the harpoon (about the weight of one hundred paper clips). This choice represents an arbitrary starting point that can be changed later if necessary. The students next determine the spring constant of their rubber band. Several methods exist for measuring k. Tina and Juan decide to hang weights of various known values on the rubber band and measure its displacement, as shown in Figure 4.4. This simple experiment produces the data shown in Table 4.1. Of particular interest are the ratios of the force F to the displacement y. These ratios are equivalent to the spring constant of the rubber band, because $k = -F/y$. They rightfully ignore the values produced by the largest weights, because the rubber band is stretched beyond its elastic limit for these very large excursions. The values in Table 4.1 lying below the elastic limit yield an average of about 75 N/m. The data also suggest that Tina and Juan should confine their stretching excursions of the rubber band to about 7 cm or less. At this point, their calculation parameters consist of the values shown in Table 4.2.

TABLE 4-1   Force–Displacement Test Data for Rubber Band

WEIGHT (gm)	FORCE |mg| (newtons)*	STRETCH y (cm)	F/Y
50	0.49	0.7	70
100	0.98	1.3	75
150	1.5	2.0	75
200	2.0	2.6	77
250	2.5	3.3	76
300	3.0	4.1	73
350	3.4	4.5	76
400	3.9	5.2	75
500	4.9	6.5	75
600	5.9	7.8	76
700	6.9	9.1	76°°
800	7.8	9.3	84
900	8.8	9.4	94
1000	9.8	9.4	104

°$F = mg$, where the gravitational constant $g = 9.8$ N/kg.
°° The probable elastic limit lies just above this point.

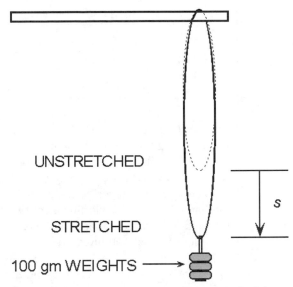

Figure 4.4. Measuring the elastic constant k of the rubber band.

The program outlined in the flowchart of Figure 4.3 begins by setting all fixed parameters and constants to those listed in Table 4.2 and by drawing a profile of the ramp on the computer screen. The program next proceeds to a pause point where it prompts

Figure 4.3. Flowchart illustrating the program used to plot harpoon trajectories. The program prompts the user to enter values for stretch length s and inclination angle θ.

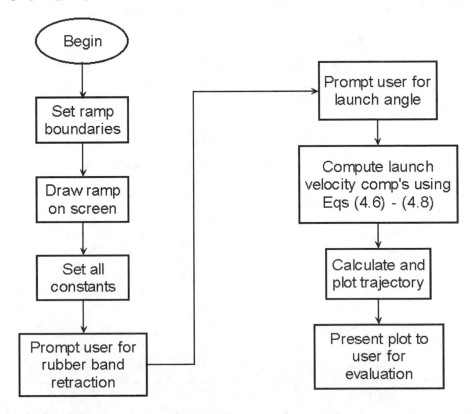

TABLE 4-2   Calculation Parameters for Numerical Simulation

$k = 75$ N/m	elastic constant of rubber band
$m = 0.1$ kg	mass of harpoon
$g = 9.8$ N/kg	gravitational acceleration
$s_{max} = 7$ cm	maximum retraction distance for rubber band

the user to enter values for the spring retraction distance s and the launch angle θ. After receiving suggested values, the program checks to make sure that $s < s_{max}$ and then calculates the exit velocity of the harpoon using Eq. (4.6) and the x- and y-components of the velocity using Eqs. (4.7) and (4.8). The program next increments the time and plots the newly calculated positions x and y of the harpoon using Eqs. (4.1) and (4.2). The time increment dt used for plotting is set to 1/100 of the horizontal span of the ramp divided by the horizontal component of the exit velocity. This choice for dt is arbitrary but leads to a plot of about one hundred points to form a smooth-looking trajectory on the computer screen.

The best values for angle θ and band stretch s are found by trial and error. Figure 4.5 illustrates a sampling of trajectories for the test values $s = 7$ cm and $\theta = 50°$, $60°$, and $70°$. The simulation indicates that a launch angle of $50°$ will cause the harpoon to collide with the uphill side of the ramp without clearing the top. A launch angle of $60°$ will allow the harpoon to pass over the ramp completely but also will cause it to land at the opponent's starting point. A launch angle of $70°$ will land the harpoon near the desired spot.

Figure 4.5.   Several calculated trajectories for various values of launch angle θ at a stretch distance of 7 cm.

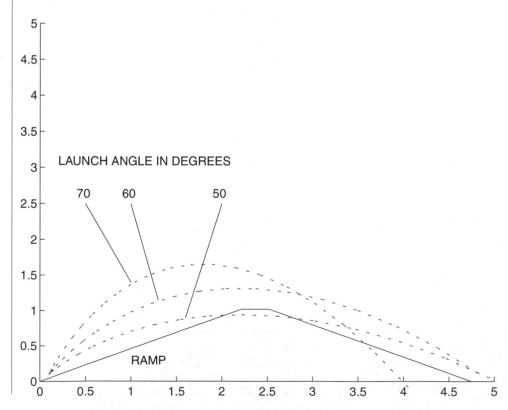

The program outlined by the flowchart shown in Figure 4.3 can be implemented using any of several programming languages or software packages, including C, Pascal, Fortran, BASIC, Mathematica, Mathcad, and MATLAB. In the next section, one possible program for implementing the simulation in the MATLAB programming language is illustrated. The problem also could be analyzed using C, Mathematica, or Mathcad.

EXAMPLE 4.2: PLOTTING TRAJECTORIES USING MATLAB

MATLAB is a versatile and comprehensive programming environment particularly suited for engineering problems. MATLAB is similar in syntax to the programming language C, but it also provides commands that allow you to quickly plot and organize data, manipulate matrices, observe variables, and solve systems of linear equations. The strength of MATLAB lies in its ability to manipulate large amounts of data while allowing these data to be plotted and graphed. Tips on getting started in MATLAB can be found in any of several references. One of the modules of the engineering series published by Prentice Hall, of which this book is a member, for example, outlines the use of MATLAB in considerable detail.

In this example, we illustrate how the trajectory calculations described by the flowchart of Figure 4.3 can be performed easily using MATLAB. One possible version of the code is shown below. Comment lines are preceded by percent signs (%). The program first plots a proportionately correct profile of the ramp, then prompts the user for values of the launch tube inclination angle and the amount of stretch of the rubber band. The program increments t by dt for each pass through the `while` loop. The looping continues as long as x and y lie within the boundaries $y > 0$ and $x < H$. After each position calculation, the program extends the trajectory plot from the most recently calculated (x,y) position to the newly calculated position. Visual inspection of the plot reveals whether the trajectory hits the ramp at the desired location. This program was, in fact, the one used by the author to produce the plots of Figure 4.5.

MATLAB PROGRAM CODE

```
H=5;V=2;          %set boundaries of the problem in meters
axis([0 H 0 H]);
hold on
%set axes for making plot on screen
%Draw ramp on the screen:
plot([0 2.22 2.52 4.75],[0 1.02 1.02 0]);
k=75;             %elastic constant of rubber band in N/m
smax=7;           %maximum allowed s in cm
m=0.1;            %mass of harpoon in kg
g=9.8;            %gravitational acceleration in m/s^2
s=input('ENTER AMOUNT OF RUBBER BAND STRETCH IN cm: ');
if (s>smax) disp('Value of s must be less than 7 cm');
keyboard; end
angle=input('ENTER ANGLE OF LAUNCH in DEGREES: ');
s=s/100           %convert stretch from cm to meters
angle=angle*pi/180;%convert angle to radians
vo=sqrt(k/m)*s;   %compute initial velocity
vox=vo*cos(angle);%compute x-component of initial velocity
voy=vo*sin(angle);%compute y-component of initial velocity
dt=H/(vox*100);   %specify a time increment
x=0.1;            %set initial values of x and y
y=0.1;
t=0;              %set time to zero
```

```
while (y>0 & x<H) %calculate until trajectory goes out of
bounds
    t= t+dt;        %increment the time
    xnew = x + vox*t; %compute new position x
    ynew= y +voy*t -0.5*g*(t^2);%compute new position y
    plot([x xnew],[y ynew]); drawnow
    x=xnew; y=ynew
end
```

PROFESSIONAL SUCCESS: GARBAGE IN, GARBAGE OUT

The computer has become an indispensable tool for the engineer. It can perform calculations much more rapidly than can a human. It's superb for storing and retrieving data and producing graphical images. But the computer is no substitute for thinking. A computer should be used to amplify the capabilities of an engineer, not replace them. A computer will faithfully follow its own program code but is unable to pass judgment on the worth of its own program. It's up to the engineer to provide the computer with information and program statements that are meaningful and relevant. If we program a computer to calculate the weight of structural steel needed to build a bridge, it can do so flawlessly.

But it never can tell us whether such a bridge is feasible, whether the stress/strain equations are correct, or whether the bridge should be built at all. Only an engineer with experience and good judgment can make those determinations. "Garbage in, garbage out" (GIGO) refers to a situation where bad input data or bad program code lead to computationally correct but meaningless output. GIGO can be prevented by testing a program on simple, well-known examples that are easily calculated by hand. If the program can provide the correct answers to simple problems, it's ready to be used on more complex problems.

Exercises 4.2

E4. Compute by hand the trajectory of Figure 4.5 for a 60-degree launch angle. Verify that the projectile lands 5 meters from its starting point on the other side of the ramp.

E5. A rubber band has an elastic constant of 100 N/m. How much force will be required to elongate the band by 5 cm?

E6. What is the potential energy stored in a spring that has a restoring force of 1 kN/m and has been stretched by 1 cm?

E7. What is the potential energy stored in a spring that has a restoring force of 500 gm/mm and has been compressed by 5 mm?

E8. Compute the initial velocity of a 100-gm projectile that is launched by stretching a rubber band by 10 cm. The rubber band has an elastic constant of 50 N/m.

E9. What launch angle will result in a harpoon that travels the farthest distance in the Peak Performance Design Competition? Assume that the harpoon's launch velocity and angle will cause it to travel over the top of the ramp and land somewhere on the other side.

E10. Draw the flowchart of a program designed to compute the path of a bowling ball after it leaves a bowler's hand.

E11. Draw the flowchart of a program designed to compute the height of a helium-filled balloon after it has been released from sea level.

4.3 SPREADSHEETS AND DATA BASES

A spreadsheet is a programmable table in which each cell consists of text, fixed numerical data, or a formula that uses other numbers in the spreadsheet. Popular spreadsheets include Microsoft Excel™ and Lotus 1-2-3™. Engineers may use spreadsheets in all phases of the design process for such tasks as planning budgets, tracking parts lists, analyzing results, and performing calculations. A spreadsheet becomes particularly useful when a problem is complex and has many interrelated variables. By programming a spreadsheet to model a complex problem, an engineer can see the effect on the entire system of changing a single variable. The following two examples, written in the context of the Peak-Performance Design Competition of Chapter 2, illustrate the usefulness of spreadsheets in engineering design. The spreadsheet tables shown are meant to reflect a generic form typical of most commercial spreadsheet software.

EXAMPLE 4.3: CALCULATING THE CENTER OF MASS

Chapter 2 described the design goals of two students, Tina and Juan, who wish to enter the wedge-shaped vehicle depicted in Figure 2.9, including launch tube, into the Peak-Performance Design Competition. They have determined that the car will perform best if its center of mass is located midway between the front and rear axles. Their simulations have shown that if the center of mass is too far forward, the rear wheels will not maintain enough traction to get up the hill. Conversely, if the center of mass is too far toward the rear, the high torque of the motor and gear drive may cause the front end of the car to lift off the ramp during startup or during contact with an opposing car. The students are now addressing the problem of where to mount the various components on the vehicle chassis. The location the vehicle components will determine the location of the center of mass.

Tina has weighed each component and has entered her list, which is shown in Figure 4.6, into her logbook. She's also entered the sketch shown in Figure 4.7, which outlines one possible layout for the components. She's calculated the center of mass of the wedged-shape frame from basic principles and has found that it lies about 7 cm from the center of its own x-axis. (See problem 4.19.)

Their next step is to decide where to place each of the components on the vehicle chassis. Juan prepares a spreadsheet, shown in Table 4.3, to calculate the center of mass of the entire vehicle, including all its components. Note that Table 4.3 shows the contents of each cell in the spreadsheet, and not what appears on the computer screen. The location x of each part represents the position of its own individual center of mass relative to the midpoint between the axles, and the moment M of the part is equal to its position times its mass m:

$$M = x \times m$$

The center of mass x_{CM} of the entire ensemble of parts is given by the sum of the moments divided by the sum of the masses:

$$x_{CM} = \frac{\Sigma\,M}{\Sigma\,m}$$

The output of Juan's spreadsheet, as it appears on his computer screen, is shown in Table 4.4. The numbers computed by the various cell formulas are shown in bold.

Juan's calculations show that the center of mass of the vehicle (Net C.O.M. in the spreadsheet table) will lie about 5 cm behind the geometrical center of the axle mounts. The force calculations that he and Tina have previously performed using a C++ program indicate that the center of mass needs to be within 2 cm of the vehicle's midpoint

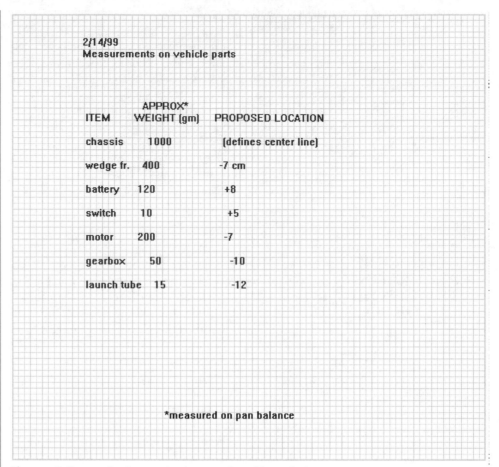

Figure 4.6. Logbook page showing weights of key vehicle components.

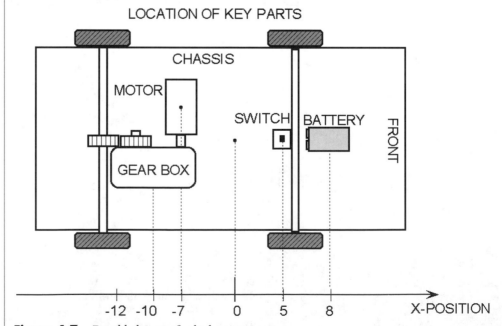

Figure 4.7. Possible layout of vehicle components.

TABLE 4-3   Cell Entries in Spreadsheet that Calculates the Center of Mass

		1		2		3		4
A		Part Name		Mass m (gm)		Location x (cm)		Moment M (gm°cm)
B		Chassis		1000		0		B2 ° B3
C		Wedge Frame		400		–7.1		C2 ° C3
D		Battery		120		8.0		D2 ° D3
E		Switch		10		–5.0		E2 ° E3
F		Motor		200		–7.0		F2 ° F3
G		Gearbox		50		–10.0		G2 ° G3
H		Launch Tube		15		–12.0		H2 ° H3
I								
J		Total Weight		sum (B2 ... H2)				
K		Total Moment						sum (B4 ... H4)
L		Net C.O.M.				(K4)/(J2)		
M								

TABLE 4-4   Screen Output of the Spreadsheet of Table 4.3

		1		2		3		4
A		Part Name		Mass m (gm)		Location x (cm)		Moment M (gm°cm)
B		Chassis		1000		0		**0**
C		Wedge Frame		400		–7.1		**–2828**
D		Battery		120		8.0		**960**
E		Switch		10		–5.0		**–50**
F		Motor		200		–7.0		**–1400**
G		Gearbox		50		–10.0		**–500**
H		Launch Tube		15		–12.0		**–180**
I								
J		Total Weight		**1795**				
K		Total Moment						**–3998**
L		Net C.O.M.				**–5.03**		
M								

if the proper balance of traction and stability is to be achieved. The present location of the center of mass is too far behind the midpoint.

The students discuss possible alternative arrangements for the vehicle components. Juan tries to simulate moving the battery forward by changing the entry for battery position on his spreadsheet. Because the battery is connected to the motor by wires of arbitrary length, this change is an easy one to implement. The change, however, shows that moving the battery all the way to the 15-cm forward position merely shifts the center of mass forward to about –4 cm.

Tina suggests adding a counterweight somewhere between the midpoint of the vehicle and its front axle. Juan inserts a new row into the spreadsheet and enters data for the counterweight into the cells, as shown in Table 4.5. The students experiment with different values for the location and mass of the counterweight. The spreadsheet allows them to track changes in the center of mass as it's modified by the counterweight. In addition, the Peak-Performance Design Competition rules

TABLE 4-5   Screen Output of the Modified Spreadsheet

		1		2		3		4
A		Part Name		Mass m (gm)		Location x (cm)		Moment M (gm°cm)
B		Chassis		1000		0		**0**
C		Wedge Frame		400		–7.1		**–2828**
D		Battery		120		8.0		**960**
E		Switch		10		–5.0		**–50**
F		Motor		200		–7.0		**–1400**
G		Gearbox		50		–10.0		**–500**
H		Launch Tube		15		–12.0		**–180**
I		**Counterweight**		200		10.0		**2000**
J								
K		Total Weight		**1995**				
L		Total Moment						**–1998**
M		Net C.O.M.				**–2.01**		
N								

specify a 2-kg limit. The spreadsheet allows the students to see how close they come to this limit as the mass of the counterweight is increased. The cells in Table 4.5 show the results of placing a 200-gm counterweight 10 cm forward of the vehicle midpoint. The center of mass has been shifted to –2 cm behind the midpoint, which lies just on the edge of the targeted range of ±2 cm. However, the total weight has increased to 1995 gm, which is just five grams short of the maximum 2-kg limit. Tina and Juan decide to stay with these parameters so that they will have a small safety margin with respect to weight, in case they need to add small nuts, bolts, glue, or tape during the competition.

EXAMPLE 4.4: KEEPING TRACK OF COST WITH A SPREADSHEET

A spreadsheet also helps Juan and Tina keep track of the total cost of their vehicle. Their professor has set a total cost limit of $100, including batteries, so that no one can produce a better vehicle simply by having unlimited funds. An accounting of the costs must be submitted by each participating team, and Tina has set up a spreadsheet to track these costs. The output screen of the spreadsheet is shown in Table 4.6. The *Unit Cost* column indicates the cost per item, and the *Quantity* column indicates the number of parts of that type that has gone into the vehicle. The *extended* column, equal to the product *Quantity × Unit Cost,* shows the total cost for each part type, and the TOTAL entry at the bottom provides the sum of all the extended prices. As each part is added to the vehicle, the students update their spreadsheet. If the total cost exceeds $100, they can use the spreadsheet to experiment with eliminating various parts, such as extra nuts and bolts, if the design allows it, until the total cost is in compliance. At first they allot sixteen batteries to the vehicle for competition day, but this quantity causes the total coast to exceed $100. By using the spreadsheet, they are able to determine by trial and error that reducing the battery allotment to twelve results in a total cost less than $100.

TABLE 4-6 Cost-Tracking Spreadsheet. The *Unit Cost* Column Indicates the Cost Per Item, and the *Extended* Column is Equal to *Quantity* × *Unit Cost*.

		1		2		3		4		
A		Item		Quantity		Unit Cost $				Extended $
B		Chassis		1		12.50				12.50
C		Wedge Frame		1		15.00				15.00
D		Battery		12		2.19				26.28
E		Switch		3		2.29				6.87
F		Motor		1		3.49				3.49
G		Gearbox		1		5.99				5.99
H		Launch Tube		1		0.67				0.67
I		Counterweight		1		0.85				0.85
J		Rubber Bands		12		0.04				0.48
K		6-32 Screws		24		0.06				1.44
L		6-32 Nuts		24		0.05				1.20
M		6-32 Washers		24		0.02				0.48
N		Two-Part Epoxy Tube		1		2.29				2.29
O		Wheels		4		1.59				6.36
P										
Q		Spool of Wire		1		2.49				2.49
R		Tape		1		2.59				2.59
S		Metal Brackets		6		1.19				7.14
T		Screw Thread		1		0.99				0.99
U		Wing Nut		2		0.25				0.50
V										
W		TOTAL								97.61
X										

Exercises 4.3

E12. Verify the calculated cells in the spreadsheet table of Table 4.4.

E13. Find the center of mass in the x-y plane of a set of objects positioned as follows: Object 1: 1.2 kg (0.2 m, 0.4 m); Object 2: 3.3 kg (1.3 m, 0 m); Object 3: 0.9 kg (0.8 m, 0.4 m); Object 4: 0.2 kg (0 m, 1.7 m);

4.4 GRAPHICAL PROGRAMMING

A category of software tool called *graphical programming* recently has become extremely popular among design engineers. Marketed under commercial software packages, such as LabView™, Simulink™, and HP-VEE™, graphical programming languages allow engineers to create programs simply by connecting together visual objects on the computer screen. An object can be a formula, a data source, a display, or a logical function. Objects are tied together much like the boxes of a flowchart to create computer programs. Some graphical programming languages allow the user to interface directly with benchtop instruments, such as multimeters, function generators, and oscilloscopes.

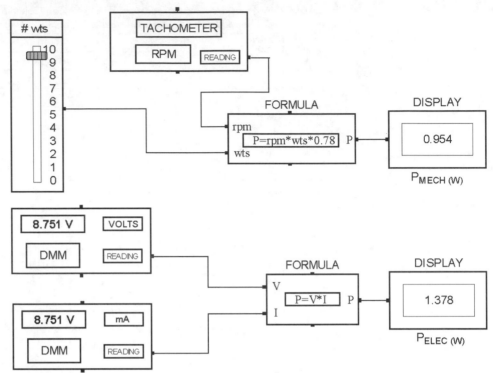

Figure 4.8.   Graphical interface program designed to measure the voltage across and current through a motor under test. The mechanical load, expressed in terms of the number of weights, is entered into the program by the user.

Interconnections are made using digital control links, such as the IEEE-488 bus, GPIB (general purpose instrument bus), and HP-IB (Hewlett-Packard instrument bus). A special bus card is installed in the computer, and cables are fed to each of the benchtop instruments. Data can be sent to and from the instruments via graphical program objects. Connections can also be made to analog-to-digital (A/D) and digital-to-analog (D/A) plug-in cards directly from the computer bus. This environment is ideal for creating automated control-and-instrumentation applications. Many industrial systems consist of A/D and D/A boards plus a graphical programming interface run from a desktop PC.

An example of a typical graphical program is shown in Figure 4.8. This program is designed to measure the voltage across and current through a motor being tested using the Prony brake apparatus of Figure 2.13. The mechanical load, expressed in terms of the number of weights, is entered into the program by the user.

4.5 THE WORLD WIDE WEB

It's almost impossible to be an engineer today and not make use of the World Wide Web. The graphical linking of millions of host sites on the Internet has become an important tool not only for engineers, but also for individuals involved in business, commerce, science, politics, education, and recreation. While a comprehensive introduction to the World Wide Web is beyond the scope of this text, a discussion of when to use the Internet as part of the engineering design process is certainly appropriate.

Most companies who make parts and supplies for engineering projects maintain Web sites that supply information about company products. Two battery-related Web sites (www.duracell.com and www.eveready.com), for example, were mentioned in Chapter 3 as the source for technical data on batteries for Peak-Performance vehicles. Other actual Web sites, including www.motionshop.com (motors and gears), contain information pertinent to the competition.

"I Saw It on the Internet. It Must be True."

When you access information from the Internet for an engineering project, be sure that it comes from a reliable source. Information is not necessarily accurate just because it has been posted on the Web. If the data comes from the Web site of a reputable company, it is most likely reliable. Information from student Web pages, project sites, the informal press, lone information providers, etc., should possibly be viewed with skepticism until it can be verified.

If a particular product or system interests you, take the time to obtain full product information. What is posted on the Web is often less comprehensive than information available in printed form for serious users. The latter usually can be obtained directly from the manufacturer by sending e-mail from the Web site.

One other word of caution: The World Wide Web has been around only since the early 1990's, and although its growth has been explosive, the information it can provide on a particular subject is only complete if someone has taken the time to post it. A great body of engineering data and knowledge exists that was present long before the Web came into being. As a source of information, the Web is no match for the hundreds of millions of books, reports, and periodicals available in the world's libraries. Although the Web and its associated computer-based searching tools can be important sources of information for your design project, they should not be the only source.

PROFESSIONAL SUCCESS: PROVIDING GOOD SOFTWARE DOCUMENTATION

When writing a complex piece of software, get in the habit of adding a comment to every line of the program code. Your comment should explain the purpose of the programming step. This procedure may seem like overkill and a nuisance while you're developing your program, but you'll be thankful you took the time should you return to your program at some later date. You'll be surprised at how quickly you can forget the subtle nuances of a program's flow logic when you're not immersed in it on a daily basis. And other engineers who need to modify or debug your code also will thank you.

Exercises 4.5

E14. Use the search feature of your Web browser to perform a search on the term "oil platform." Record the number of listings. Now narrow the search by adding in succession the addition keywords "photo", "north", "sea", and "terminal." Determine how many fewer listings are obtained each time the search is narrowed by an additional keyword.

E15. Use the search feature of your Web browser to perform a search on the term "catapult" Record the number of listings. Now narrow the search by adding in succession the addition keywords "photo", "medieval", "trebuchet", and "reenactment." Determine how many fewer listings are obtained each time the search is narrowed by an additional keyword.

E16. Use the search feature of your Web browser to perform a search on the term "space station." Record the number of listings. Now narrow the search by adding in succession the addition keywords "photo", "NASA", "Russian", and "shuttle." Determine how many fewer listings are obtained each time the search is narrowed by an additional keyword.

4.6 REAL-TIME COMPUTER CONTROL

We usually associate the term "computer" with the desktop PC or networked workstation. In truth, these integrated computational machines represent but a small fraction of the all the computers in use today. The most prolific computer, the *microprocessor,* far exceeds in numbers the desktop PC. The *microprocessor* is a single-chip computer that performs digital-logic functions at the fundamental level. Microprocessors are excellent choices for solving many design problems, especially those involving real-time control by computer. These applications are sometimes referred to as embedded computing.

In its most basic form, a microprocessor has no disk drive, keyboard, monitor, or external memory chips, but simply consists of a silicon microcircuit housed in a plastic or ceramic package. A microprocessor lies at the heart of just about every appliance or piece of equipment that requires intelligent control. Microprocessors can be found inside automobiles, microwave ovens, washing machines, children's toys, cellular telephones, fax machines, printers, and, of course, personal computers. The Pentium® chip, made by Intel, and the PowerPC® chip, made jointly by Apple™ Computer and IBM™ are examples of high-performance microprocessors that are connected to peripheral devices to create desktop PCs. The more numerous microprocessors found inside appliances and machines are usually much simpler devices than Pentium® or PowerPC® chips, but their operating principles are the same. Microprocessors operate in a base-two arithmetic system of logic rules known as *Boolean algebra.* The rules of Boolean algebra allow a microprocessor to make decisions based on the status of its stored or incoming digital data. The collection of these elementary operations comprises a machine-level computer language called *assembly code.* The typical language of assembly includes such low-level instructions as *read port, store in memory, add, subtract, shift left, shift right*, and *compare byte.* Assembly language also provides simple branching and testing instructions, (e.g., "if . . . then" or "branch if equal" commands). More complex operations, such as floating-point calculations, data manipulation, data communication, and control, are performed by piecing together, or *assembling,* sets of the more basic instructions. A microprocessor executes its machine-level instructions extremely rapidly and is ideal for real-time applications.

Simple programs can be created quickly by writing code directly in assembly language. More complex programs are typically written in higher level programming languages, such as C, C++, Pascal, or even BASIC. A special program called a *compiler* reads the high-level program and writes appropriate assembly language code. The machine-level assembly code instructions are then transferred, or *downloaded,* to the microprocessor and stored in its permanent, read-only memory (ROM). In the design stage, when a program is being developed, a special, more expensive version of the microprocessor chip is used that permits the code to be erased and rewritten. The erase procedure requires that the microprocessor be exposed to ultraviolet light for several minutes. The chip package contains a special transparent window for this purpose. When the final design of the microprocessor-controlled device is massed produced, less expensive versions of the microprocessor that allow only a one-time burn-in of the assembly code into ROM are used.

Microprocessors communicate with the outside world via sets of terminals, called *ports,* that carry digital signals. A digital signal is a voltage that is either low (0 V), representing the binary number zero, or high (3 V or 5 V, depending on the system), representing the binary number one. A single port consists of a group of 8, 16, 32, or even 64 terminals activated by the microprocessor. A port is bidirectional—it can bring digital data into the microprocessor or send digital data out from the microprocessor. The job of the microprocessor is to perform basic arithmetic or logical operations on data before sending the results to the outside world. Many microprocessors also include an internal *timer* that allows the device to keep track of time.

4.7 ANALOG-TO-DIGITAL AND DIGITAL-TO-ANALOG CONVERSION

Sometimes microprocessors communicate with the outside world using additional circuits called *analog-to-digital* (A/D) and *digital-to-analog* (D/A) converters. Most physical quantities, such as temperature, pressure, velocity, and position, are analog quantities. An analog variable has meaning over the entire span of its available range. In electronic systems, analog quantities are typically represented by voltage signals that vary between two limiting values, such as 0 V and 5 V. Sensors have the job of converting a real physical variable, such as position or temperature, into its equivalent analog voltage representation. Suppose, for example, that a sensor produces an analog voltage that represents the position of a sliding bracket. A value of 5 V corresponds to the right-most position excursion at 1 m, and a value of 0 V corresponds to the left-most excursion at −1 m. If the bracket lies at a position 0.5 m from the right, the analog sensor monitoring its position will produce a voltage of 3.75 V, or three-fourths of the voltage range between 0 V and 5 V.

In contrast to an analog voltage, a digital voltage, or *bit,* can take on only two states that have meaning. The on state of a digital signal corresponds to the binary number **1,** and the off state corresponds to the binary number **0.** A digital electronic system usually adopts a voltage level near 0 V to represent binary **0** and a voltage level near either 5 V or 3 V, depending on the system, to represent the binary number **1.** Joining together several digital bits produces a multibit binary number. For example, a collection of eight bits can be used to represent any number between **0000 0000** (zero) and **1111 1111** (255 in base ten.)

An A/D converter accepts an analog voltage as input, compares the incoming voltage to a reference voltage, and produces an equivalent multibit binary number that represents the ratio of the input to the reference voltage. As an example, suppose that the A/D converter compares its input to a 5-V reference. If its input is equal to 2.5 V and the conversion is based on eight digital bits, then the converter will represent the input by the digital output **1000 0000,** or 128 in base ten. This output represents 128/255ths, or approximately half, of the analog range, indicating that the ratio of the analog input to the reference is 2.5/5 = 0.5. An input of 1 V to the same converter will yield an output of **0011 0011,** or 51 in base ten. This output represents 51/255ths, or approximately 20 percent, of the analog range, indicating that the ratio of the analog input to the reference is 1/5 = 0.2.

If a microprocessor sends out a digital signal, it can be converted to analog form by a D/A converter. A D/A converter takes a multibit binary signal as its input and produces an analog voltage as its output. The availability of A/D and D/A converters greatly enhances the utility of a microprocessor and allows it to communicate with the real physical world. As illustrated by the next example, microprocessors are used frequently by computer and electrical engineers to help solve design problems.

Exercises 4.6

E17. Convert each of the following binary numbers to decimal (base ten): **0001, 1000, 1111, 1010.**

E18. Write the binary (base two) presentation of each of the base-ten digits from 0 through 9.

E19. An analog-to-digital converter has four bits of resolution, and its reference voltage is 10 V. What voltage increment is represented by each binary digit?

E20. A digital-to-analog converter has four bits of resolution, and its reference voltage is 5 V. What output voltage is produced by the inputs **1000** and **1111?**

E21. How many wires does it take to send a four-bit digital signal from one circuit to another?

E22. Draw the serial data stream for the bit sequence **1011.**

EXAMPLE 4.5: MICRO-PROCESSOR SPEED CONROL FOR PEAK-PERFORMANCE VEHICLE

One interesting approach to the Peak-Performance Design Competition introduced in Chapter 2 involves the use of a microprocessor to determine the speed and position of the vehicle. This approach allows for precise control of the stopping point at the top of the hill. Suppose that a sensor were available that could provide information about the net distance traveled by the vehicle. One form of such a sensor is depicted in Figure 4.9. A transparent disk with eight opaque radial lines passes between the arms of an optical detector. The disk is connected to an idler wheel much like that shown in Figure 3.10. After each 22.5° of rotation of the transparent disk, a line falls between the arms of the optical sensor causing it to send a pulse to the microprocessor. Because the circumference of the idler wheel is known, the distance traveled by the car between pulses is also known. By simply counting pulses, the microprocessor can compute the total distance traveled by the vehicle, and the microprocessor can issue a stop command when it determines that the car has reached the top of the hill. This design approach represents a marked improvement over the open-loop timing scheme and the sliding-nut idea presented in Chapter 3. For one thing, the system will not be sensitive to motor speed or battery voltage. Moreover, an idler wheel driving a simple transparent disk that produces almost no friction will be much less likely to slip or drag along the ramp than will a wheel required to turn a threaded rod or move a sliding nut, as depicted in Figure 3.10. The sliding-nut system of Chapter 3 will have much more mechanical friction to overcome than will the optical-disk system shown in Figure 4.9.

Using its internal timer, the microprocessor also will be able to determine the car's instantaneous velocity by computing and constantly updating the ratio of distance traveled to elapsed time. Likewise, the microprocessor will have the capability of setting the speed of the car to a value specified by the user at the start of each run. The desired speed would be set via a set of small switches (say, three in number) that are moved to

Figure 4.9. Optical encoder wheel produces a digital pulse every 22.5° of rotation.

OPTICAL SENSOR

TRANSPARENT WHEEL WITH OPAQUE LINES

on or off positions representing logic **1** and **0,** respectively, thereby forming one of $2^3 = 8$ binary numbers lying between **000** and **111.** The microprocessor would first read the switches to determine the desired speed setting, then drive the motor at a level appropriate for that velocity via a digital-to-analog converter. Adopting this clever design would allow the user to choose between a selection of eight preset velocities before each run up the ramp. A slow speed, for example, might be chosen as a defense against a wedge-shaped car in the hope that arriving second at the top of the hill would result in ultimate dominance. Conversely, a fast speed might be chosen as a defense against a projectile system. A car that could zoom to the top of the hill ahead of an opponent's jettisoned object might have a better chance of dominating the top of the hill. A flowchart showing the data input and decision-making requirements of such an embedded microprocessor is shown in Figure 4.10.

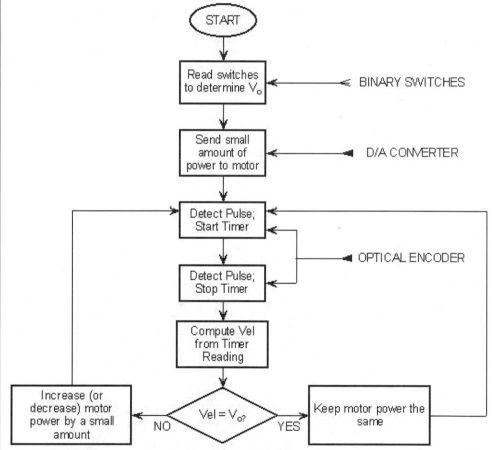

Figure 4.10.   Flowchart showing data input and decision making for the microprocessor program.

EXAMPLE 4.6: MICRO-PROCESSOR CONTROL SYSTEM FOR ELECTRIC CAR

The following example poses a microprocessor control problem similar to the previous example, but not directly related to the Peak-Performance Design Competition. Imagine an engineer who must design a microprocessor-based control system for an electric automobile (a real car, not a model). A microprocessor will serve as the main control system for the vehicle and integrate input from the driver with signals from numerous sensors. The engineer needs to design an interface between the embedded microprocessor and the analog electronic circuit that will power the vehicle's main drive motor. One of his design tasks is to decide upon the form of the signals that pass between the microprocessor and the motor-drive circuitry. The specific microprocessor, still to be chosen, will depend on the final requirements of the system, but the engineer has decided to base the design on an 8-bit A/D and D/A system. This choice will provide a resolution of about one part in 256 (i.e., $1 \div 2^8$) for any variable processed by the microprocessor.

The microprocessor module will reside beneath the car's dashboard, and the motor-drive circuitry will be mounted inside the engine compartment as close as possible to the motor. This choice of location for both circuits requires the routing of wires between them through the fire wall of the vehicle. The engineer must decide whether to send signals between the motor-drive circuit and the microprocessor in analog or digital form. This critical design decision will affect other aspects of the system, and the engineer realizes that he cannot design the motor-drive circuit until he decides whether it will interface with the microprocessor via analog or digital signals. The engineer has compiled the following list that compares the advantages and disadvantages of each method:

Comparison of Microprocessor/Motor Power Circuit Interconnection Methods

FEATURE	ANALOG SIGNAL LINK	DIGITAL SIGNAL LINK
Number of wires required per signal	One plus common gnd	Eight plus common gnd
A/D and D/A requirements	May be able to use microprocessor that has internal A/D and D/A circuitry	Needs external A/D and D/A chips located on the motor-power circuit board
Noise sources	—Electromagnetic pickup —Contact resistance —Electrical noise from motor brushes —Thermal noise	—Digital "spikes" and "glitches" —Digital crosstalk —Electrical noise from motor brushes

One important design parameter is the number of wires leading through the firewall from the microprocessor to the motor controller. Some microprocessors contain internal D/A and A/D circuits, so that only two wires per signal, one for the analog voltage and one for the return ground, will be required. (Two wires are needed to make a complete circuit.) Similarly, if the D/A and A/D conversions are performed by external integrated circuit chips located on the microprocessor's circuit board, all signals can be sent in analog form. If external A/D and D/A chips are located at the site of the motor-drive circuit module, the signals must be sent through the firewall in digital form. This method will require eight wires for each signal plus a common return ground. From the standpoint of wiring reliability only, the analog form is preferable, because fewer wires will be needed. With fewer wires installed, the chance of wire breakage or connector malfunction is reduced. The weakest point of any wiring interface is usually its connectors, and the connectors available for two-wire, analog-type connections are much more robust than those available for digital wiring. The latter are typically flat, ribbon cable connectors, such as those found inside computers. The former can be the hefty, interlocking types found in conventional automobiles.

Before deciding upon a two-wire analog connection between the motor-drive circuit and the microprocessor, the engineer must consider other factors. He realizes that by sending signals in analog form and by using a microprocessor containing internal D/A and A/D converters, the cost of external components can be eliminated. This feature will lead to a less expensive overall design. Conversely, if the signals are sent in digital form, and the D/A and A/D conversions are performed at the site of the motor-drive circuit, the required external chips will have to be added to the motor controller, adding more cost to the system. This cost factor also favors sending signals in analog form.

The engineer next thinks about reasons why he might want to send the signals in digital form. For one thing, digital signals will be much less susceptible to electronic *noise* in the form of random, unwanted voltage fluctuations caused by extraneous sources. The *signal-to-noise ratio* of a system describes the ratio of desired signal to undesirable noise. The higher the signal-to-noise ratio in an interconnected system, the more chance exists that the circuit design will be successful. Sources of analog noise include electromagnetic interference from lights, appliances, power cables, spark plugs, motor brushes, and other devices that switch current. Noise affecting analog signals can also be produced by changes in resistance due to oxidation or dirt at connection points. Thermal noise can be generated by the components themselves, but this latter form of noise should be small enough that it can be ignored in this particular application. Fortunately, the electric car will have no spark plugs to generate sparking noise. On the other hand, the main drive motor will have brushes and a commutator that are notorious sources of electrical noise. The brush contacts bringing current to the rotating armature of the motor will generate considerable noise in the form of electromagnetic interference as the motor turns.

While these various sources of noise can add extraneous voltage components to analog signals, they also can affect digital signals. But in the digital case, the noise level must be on the order of several volts in magnitude before it can adversely affect the circuit's interpretation of its 0-V and 5-V digital signals. For this reason, extraneous noise voltages matter much less in a digital system. On the other hand, if the physical circuit is not designed properly, the digital signals themselves may generate ringing, or voltage spikes, caused by high-speed switching at the microprocessor's output ports. Voltage spikes can occur whenever wires carrying signals behave more like capacitors and inductors than as simple wires. A digital interface also must contend with *crosstalk*, or the tendency of digital signals on one wire to couple to other, nearby wires. Crosstalk can occur when a digital voltage undergoes a transition between its low and high values.

The engineer decides upon a digital link between the microprocessor and motor controller. While recognizing that no solution will be perfect, he reasons that he can design the circuit properly to minimize all the sources of digital noise previously mentioned, but can't as easily guarantee the noise integrity of an analog system. He chooses to send the signals in digital form, even though he'll need more wiring.

As he begins work on his interface, he realizes that a simple modification could provide the best of both worlds. If he sends the 8-bit digital data in *parallel* form, he will have to send all eight bits at once over eight separate wires plus a common ground. Alternatively, he can send the data in *serial* form, one bit at a time, over *one* wire plus ground. If the signal is sent in serial form, timers on both sides of the data link must keep track of when the bits arrive. If both sides of the link time the bit arrivals independently, the link will be an *asynchronous* link. If common bit timing is maintained via a timing signal sent over a second signal wire, the link will be *synchronous*. Because data bits are sent sequentially, serial links are usually much slower than parallel links. Similarly, asynchronous serial links are usually slower than synchronous serial links, because extra time is needed to manage bit timing. The electric-car microprocessor application does not have particularly high-speed data requirements, so an asynchronous serial link should, in principle, work just fine. Many microprocessors contain the circuitry needed to convert the eight bits of parallel digital data into serial form, ready to transmit. Only the motor board will require the extra chips needed to convert the serial signal back to eight parallel bits ready for D/A conversion. The engineer modifies his design to incorporate an asynchronous serial link of digital data with D/A and A/D conversion performed on the motor controller board. This choice provides the best compromise between connection reliability and noise immunity.

EXAMPLE 4.7: SOFTWARE FOR REAL-TIME CONTROL

The previous example dealt with the design of a communication link between the microprocessor control module and the motor-drive circuit of an electric car. In this example, we discuss the requirements of the microprocessor *software* that will send the actual commands to the motor controller. The program inside the microprocessor must digest all available information and decide whether to accelerate, brake, freewheel (allow the car to coast), or cruise (maintain constant speed by placing the car into a feedback control mode). The microprocessor will have access to several binary signals— either *true* (logic **1**) or *false* (logic **0**)—that indicate the status of the car's ignition key, accelerator pedal, brake pedal, and clutch pedal. A true (logic **1**) signal will indicate a depressed pedal or activated switch. A false (logic **0**) signal will signify a pedal or switch that has not been pressed. Analog signals, converted to 8-bit binary numbers by analog-to-digital (A/D) converter circuits, will provide information about the car's speed and degree of accelerator pedal activation. The microprocessor will use these inputs to make decisions about the car's future actions. A flowchart of the program might look something like the diagram shown in Figure 4.11.

Every time the car's start key is turned on, the microprocessor will run an initialization program from a set of assembly code instructions stored in ROM (read-only memory). Similar in function to the boot program of a desktop computer, the initialization program will prepare the processor's various registers and ports for input or output, set program flags to be used by the main software, and perform system checks. The startup sequence also will verify that all sensors are operating, that no faults exist in the brake or drivetrain systems, and that lamps and headlights are operational.

After the startup process, which will take no more than a few milliseconds, the microprocessor will begin executing the main section of its software program. The program first will query the various pedals to determine their status. If the brake is pressed with the car at rest, no power will be applied to the motor and the car will remain at rest. Similarly, if the car is moving, but neither the brake nor accelerator is pressed, the program will assume that freewheel mode is desired and again apply no power to the motor. If the cruise lever is engaged, the program will check the car's speed, record its value temporarily in random-access memory (RAM), and return to the pedal-check point in the program. If the pedal status has not changed since the last iteration, and if the cruise lever is still engaged, the program will loop again to the cruise-mode branch, where it will recheck the speed and compare it with the value stored in RAM. If a speed decrease has occurred due to frictional losses, the motor power will be increased slightly before the program returns to the pedal-check point. Conversely, if the speed has increased since the last speed check, the power will be reduced slightly. Over many passes of the loop, each of which takes only a fraction of a second, the car speed will be kept at a constant value while the car is in cruise mode.

If the pedal check reveals a released brake and a depressed accelerator, the program will enter acceleration mode. It will assess the car's current speed and then compute the power needed to achieve an acceleration proportional to the amount of pedal actuation before looping back to the pedal-check point.

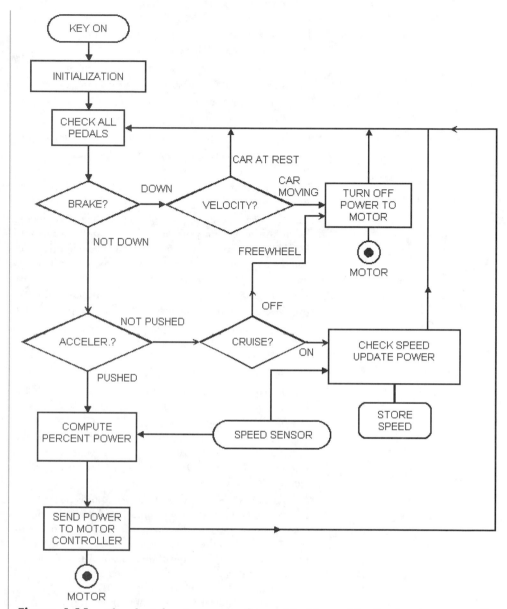

Figure 4.11. Flowchart showing program flow and decision making for electric car controller.

KEY TERMS

Computer	Software	Hardware
Analysis	Microprocessors	Control
Digital data		

Problems

Use of Computers

1. Discuss the way in which a computer or microprocessor (single-chip computer) might improve your approach to designing the following products:

     a.   An all electronic telephone answering machine

     b.   A tachometer and an odometer for a bicycle

     c.   An energy-saving light switch

     d.   A smart clothes iron that shuts off after one hour without use

     e.   A data logger for measuring part weight in a quality-control system

     f.   A voice-synthesized device for a speech impaired person

     g.   A digital alarm clock

     h.   A system for producing Gantt charts for homework assignments

2.   Identify a system or entity in your school, home, or place of work that would benefit greatly by the introduction of a computer system. Write a short summary explaining why.

3.   Identify a system or entity in your school, home, or place of work that suffers because one or more computers were introduced inappropriately. Write a short summary explaining why.

4.   Take a survey of a select group of people who use computers. (The people you choose could be those who live on your dorm floor, attend one of your classes; or are in your extended family, for example.) Make a list of how many hours per day each person uses a computer and the approximate percentage of time spent at various tasks on the computer. Examples of tasks include word processing, spreadsheet use, CAD drafting, computation, e-mail, etc.

5.   Make a list of at least ten appliances or machines that you encounter on a regular basis, exclusive of desktop computers, that utilize the power of a microprocessor.

6.   Make a list of at least five appliances or machines that you encounter on a regular basis whose function could be greatly improved by the use of a microprocessor.

The Computer as an Analysis Tool

7.   Write a computer program in the language of your choice to calculate the trajectory of a pebble that falls from an airplane traveling at 200 kph. Ignore the effects of air resistance.

8.   Consider the snapping mousetrap bale shown in Figure 2.7. Write a computer program in the language of your choice to plot the angle θ as a function of time from $t = 0$ and $\theta = \pi$ until the time when the bale makes contact with the base at $\theta = 0$. Assume that the bale has a moment of inertia 0.01 kg-m and that the spring exerts a torque of value $0.5\theta/\pi$ N-m, where θ is in radians.

9.   A capacitor is a device that stores electrical energy. The degree to which a capacitor is charged at any given moment is indicated by how much voltage appears across its terminals. If a resistor is connected across a charged capacitor, then the current i flowing out of the capacitor and into the resistor will be given by the equation $i = v/R$, where v is the capacitor's voltage and R the value of the resistor. The capacitor will respond to the flow of current by reducing its voltage according the equation, $dv/dt = i/C$. Write a computer program in the language of your choice to plot $v(t)$ for the case $v(t=0) = 10$ V, $R = 10$ kΩ, and $C = 100\ \mu$F. (Note: 10 k$\Omega \equiv 10{,}000$ ohms; 100 μF $\equiv 10^{-4}$ farads.)

10. a.   Draw a flowchart for a computer program that could be used to control the traffic at a busy intersection where two streets cross. Traffic should be allowed to flow over the east–west route, unless a car stops at the north- or south-bound streets entering the intersection.

      b.   Write this program in the language of your choice. Include an input mechanism for indicating the number of cars at each sector of the intersection.

11. a.   Draw a flowchart for a computer program that could be used to control the traffic at a busy intersection where two streets cross. Traffic should be allowed to flow over the east–west route until three cars are stopped at the north street entering the intersection, but only if no car is stopped at the south entrance. If more than three cars become stopped along the east–west route, it should be open to traffic flow, regardless of the number of cars stopped at the north–south streets.

      b.   Write this program in the language of your choice. Include an input mechanism for indicating the number of cars at each sector of the intersection.

12. a. Draw the flowchart for a computer program that can serve as a three-digit password decoder for an alarm system. Each of the digits (0–9) entered into the alarm should be represented in binary form. Choose the last three digits of your birthday year as the password.

    b. Write this program in the language of your choice. Include an input mechanism for each of the entered digits.

13. a. Draw the flowchart for a computer program that can tally the voting of a ten-person city council. The output should indicate the majority of aye or nay volts with a logic high (**1**) or logic low (**0**) output. Include a provision for a tie vote.

    b. Write this program in the language of your choice. Include an input mechanism for each of the ten city council votes.

14. Draw the flowchart and block diagram of a sensor system that can turn on a garden watering system if the temperature rises above 30°C, the sun is not shining on the plants, and the time is not before noon of the same day.

15. Draw the flowchart for a computer program that can be used by a scientific investigator to assess the probability of various events. The system should accept five input signals and provide an output that corresponds to the status of the majority of inputs.

16. Draw the flowchart of a microprocessor program that can be used to sound an alarm in a four-passenger automobile if the ignition is energized but the driver has not put on a safety belt. The alarm also should sound if a passenger is seated but has not put on a safety belt.

17. This problem illustrates the concept of *amplitude modulation*. Suppose that the function $c(t)$ is equal to a triangular waveform that has peak values of +1 and −1 and frequency f_1. Similarly, $m(t)$ is a square wave function that varies between +1 and −1 and has frequency f_2. Write a computer program that plots the amplitude-modulated waveform described by the equation $v(t) = c(t)[1 + am(t)]$ for the case $f_1 = 10$ Hz, $f_2 = 1$ Hz, and $a = 0.4$. The factor a is called the *modulation index,* and hertz (Hz) has the units of cycles per second.

18. Draw a flowchart that will implement the system described in the following memo:

    To: Xebec Design Team

    From: Harry Vigil, Project Manager

    Subject: ATM Simulator

    Our client for this project is in charge of teaching programs for mentally handicapped students in a regional school system. The curriculum followed by these students includes practicing such common tasks as calling on the telephone, going to the supermarket, making change, and taking the bus. Their teacher has requested that we develop a machine that can simulate the functions of an automatic teller machine (ATM) such as one might find at any local bank. The availability of such a system will allow the teacher to train students without tying up an actual bank ATM machine.

    Your task is to design and build such a simulator using the components and materials of your choice. The details of the simulator's operation should be fine tuned after your initial meeting with the customer; however, here is a list of specifications that you can use as a guide in preparing your initial project proposal and technical plan:

    • The simulator should be self-contained with no actual contact with the outside world.

    • It should realistically simulate such features as user inquiry, prompting for password, type of transaction, and dollar amounts.

    • The simulator should include its own set of entry keys or buttons, display device, printer, and dispenser for (simulated) money.

    • The simulator should be triggered into operation by the insertion of an ATM-type bank card. The decoding and interpretation of information stored on the inserted cards is not a necessary feature of the system. An acceptable solution could instead involve storing passwords (possibly as an updatable list) inside the simulator; this list would be activated by the insertion of any card of the appropriate size.

Spreadsheets

19. Write a spreadsheet program that computes the trajectory of the flying harpoon of Section 4.1. Your spreadsheet calculations should be modeled after Eqs. (4.1) - (4.8), and it should have cells in which you can enter key parameters of the problem such as ramp dimensions, launch angle, and amount of rubber band stretch. Your spreadsheet should yield the same results as those found in Example 4.2, as summarized by the plots of Figure 4.5. Use an array of cells to indicate the position of the harpoon at various points along its trajectory.

20. Suppose that you are trying to come up with a budget for an engineering design project. Your salary is about $4000 per month, and the technician with whom you work earns about $2500 per month. You need $5000 for materials and supplies, $1200 for traveling to a sales meeting, and you have to contribute 8 percent of the direct dollars you spend to support clerical staff. In addition, you have to pay 80 percent of the entire contract, including the 8 percent clerical charges, to overhead that supports the general operation of the company. Write a spreadsheet program that can help determine the maximum number of person-months that you'll be able to charge to the project if the total budget must not exceed $100,000. You'd like your technician to work at least half time on the project.

21. Write a spreadsheet program that will help you determine the center of gravity of all the passengers on a small commuter airplane. The airplane has ten rows of four seats each, with two seats on either side of the aisle, for a total of forty seats. The centers of the aisle seats are located 0.5 m from the aircraft centerline, and the centers of the window seats are located 1.2 m from the aircraft centerline. The rows are spaced 1 m apart. Your program should compute the center of gravity measured relative to the first row of seats (y-coordinate). Assume that you know the weight of each passenger. Feel free to do your analysis in either kilograms or pounds.

22. A film-manufacturing plant produces standard 35-mm photographic film for cameras. The film is produced on large rolls between 500 m and 2000 m long. Each large roll might have a width between 0.5 m and 2.0 m in steps of 0.5 m. That is, there are four possible values for the width. The large rolls are sliced into 35-mm strips and packaged into consumer-sized canisters for 12, 24, or 36 pictures per strip. Assume that each picture requires 35 mm of strip length, and that the 35-mm wide strip inside each canister must allocate 10 percent of its length to a trailer and leader (i.e., unexposed film at the start and end of each roll). Write a spreadsheet program that will allow you to the compute the total number of film canisters of each size obtainable from a large roll for various percent allocations of the three values of shots per canister. Your spreadsheet should have the following user entries: width of large roll, length of large roll, percent each of 12-, 24-, and 36-shot canisters desired from entire roll. Assume that the slicing process generates no wasted film.

23. Suppose that the owner of a ferry boat has asked to you design a system for helping to load cars and freight in the most balanced way. Write a spreadsheet program that will help you determine the moment of inertia about the center of gravity of the ferry due to all the passengers and freight on the ferry. The ferry is to have forty parking spaces, each 2.7 m × 4 m, and should accommodate as many 2.7 m × 8 m shipping containers as possible. The total cargo area for the ferry is 50 m × 100 m. Assume a weight of 1000 kg per vehicle and 6000 kg per shipping container.

Real-Time Computer Control

24. Convert each of the following binary numbers to decimal (base ten):

a.	**11 0000**	d.	**1010 1010**		g.	**1111 1111 0000**		
b.	**10 0110**	e.	**1110 0111**		h.	**1010 0101 1010**		
c.	**01 1110**	f.	**1100 1011**		i.	**0111 0111 0111**		

25. Write the binary (base-two) presentation of each of the following numbers:

a.	8	d.	256	g.	5061		
b.	127	e.	1001	h.	6977		
c.	255	f.	2087	i.	10,657		

26. How many binary bits does it take to represent an analog voltage to a resolution of at least 1 mV if the range of the analog voltage is 0–5 V?

27. An analog-to-digital converter has eight bits of resolution, and its reference voltage is 10 V. How large an increase in measured signal voltage is required to increment the value of the binary representation of the signal by one bit?

28. List at least ten possible sources of electrical noise that might effect a sensitive electronic circuit. Each item on your list should be something likely to be encountered in everyday life (e.g., *not* an alien spacecraft).

29. How many wires does it take to send a twelve-bit digital signal from one circuit to another?

30. Computers communicate with other computers and peripherals using either serial or parallel data links. When a parallel link is used, the connection to the computer consists of one wire for each of the bits in the digital word, a common return ground wire, plus additional wires for sending synchronizing signals. The latter are needed so that the receiving device will know when to read each digital word sent by the transmitting device.

    When a serial link is used, data bits are sent one at a time. In a *synchronous link* system, one wire is used by the transmitting device to send the data bits in sequence, one wire is used for return ground, and a third wire is used to send a synchronizing signal. The latter is used by the receiving device to determine the timing between each data bit. In an *asynchronous link* system, typical of the type used to communicate over telephone modems, long-range networks, and the Internet, only one wire pair is available for signal transmission. Data-bit synchronization requires that the sending device and receiving device both be set to the same BAUD (for *bits audio*), or bit timing rate. Such timing is never perfect, however; if left uncorrected, the BAUD timing of the receiving device will drift apart from the BAUD timing of the sending device. To ensure that the bit-timing sequences will match, the receiving device resets its timer after each digital word. It knows when the received word ends, because the transmitting word appends a *stop-bit sequence* to each one. After the stop bit is sent, the transmitting device sets the data line to the value **1** as a prelude to sending its next word. It also adds a *start bit* of value **0** to the beginning of each word so that its arrival will be unambiguous.

    a. A particular computer sends data in eight-bit packets, or *bytes*. Determine the content of the two bytes represented by the data sequence shown in Figure 4.12. The start bit, from which you can determine the time interval per bit, is shown in bold. The stop-bit sequence consists of two data bits held high.

Figure 4.12.   Asynchronous serial data bit stream.

    b. Draw the serial data stream for the byte sequence (**1001 1100**) (**0001 1111**) (**1010 1010**).

31. Indicate the type of data link (parallel, synchronous serial, or asynchronous serial) used by a PC to communicate with each of these peripherals:

    a. Parallel printer interface

    b. External 28.8 kbps modem

    c. MacIntosh printer interface

    d. Flatbed scanner

    e. Electronic piano (MIDI interface)

32. A *pulse width modulation* motor drive system applies voltage to a motor while adjusting the *duty cycle,* or time interval, over which full voltage is applied. The current that flows when voltage is applied will be determined by the motor speed and internal resistance. The average power consumed by the motor will be equal to the time average of the voltage–current product. The pulse–width modulated waveform is produced in response to a digital data signal from a computer module. Write a program in C, MATLAB, or the language of your choice that can produce the required voltage waveform. Your program should accept a binary or decimal number between 0 and 255 and then produce an output that is high (logic **1**) for an amount of time proportional to the input value.

Analog-to-Digital and Digital-to-Analog Conversion

33. Analog-to-digital interfacing is an important part of many computer-controlled engineering systems. Although most physical measurement and control involves analog variables, most data collection, information transmission, and data analyses are performed digitally. A/D and D/A circuits provide the interfaces between analog and digital worlds.

    A D/A converter produces a single analog output signal, usually a voltage, from a multibit digital input. One common conversion algorithm produces an analog output proportional to a fixed reference voltage as determined by the equation

    $$v_{OUT} = n\ V_{REF}/(2^N - 1)$$

    where N is the number of bits in the digital input word, n is the decimal value of the binary number represented by all the input bits that are set to **1** in the digital input word, and V_{REF} is a reference voltage. When n is equal to $(2N - 1)$, v_{OUT} is equal to V_{REF}.

    a. Suppose that the input to an 8-bit D/A converter is **0010 1111** with $V_{REF} = 5$ V. Find the resulting value of v_{OUT}.

    b. A 10–bit D/A converter is fed the input word **00 1001 0001** and is given a reference voltage of 5 V. What is the output of the converter?

    c. What is the smallest *increment* of analog output voltage that can be produced by a 12-bit D/A converter with a reference voltage of 10 V if the algorithm previously shown is used?

    d. What is the largest analog output that can be produced by an 8-bit D/A converter if $V_{REF} = 12$ V?

34. An A/D converter compares its analog input voltage to a fixed reference voltage and then provides a digital output word B given by

    $$B = int\ v_{IN}(2^N - 1)/V_{REF}$$

    where the operator *int* means "round to the nearest integer". This encoding operation is called *binary-weighted* encoding. A full-scale binary output (all bits set to **1**) occurs when $v_{IN} = V_{REF}$.

    a. An 8-bit binary-weighted A/D converter has a reference voltage of 5 V. Find the analog input corresponding to the binary outputs (**1111 1110**) and (**0001 0000**).

    b. Find the binary output if $v_{IN} = 1.1$ V.

    c. Find the resolution of the converter.

    d. Find the additional voltage that must be added to a 1-V analog input if the digital output is to be incremented by one bit.

35. Boolean algebra is a system of logic used by many computers. In Boolean algebra, variables take on one of two values only: TRUE, (logic **1**) or FALSE (logic **0**). Boolean operators include AND, OR, and NOT. The AND operator is represented by a dot between variables, (e.g., $Y = A \cdot B \cdot C$ means that Y is true if A, B, and C are *all* true). The OR operator is represented by plus signs (e.g., $Y = A + B + C$ means that Y is true of one or more of A, B, or C is true). The NOT operator, represented by an overbar, simply reverses the state of the variable (e.g., $\bar{A} = \mathbf{0}$ if $A = \mathbf{1}$).

    a. Verify the following equations in Boolean algebra:

    $$A \cdot B + A \cdot \bar{B} = A$$

    $$(A + B) \cdot (B + C) = B + A \cdot C$$

    $$(A + B) \cdot (\bar{A} + C) = \bar{A} + \overline{C + B}$$

    b. DeMorgan's theorem states that the Boolean expression $\overline{A \cdot B \cdot C}$ is equivalent to $\bar{A} + \bar{B} + \bar{C}$. Similarly, $\overline{A + B + C}$ is equivalent to $\bar{A} \cdot \bar{B} \cdot \bar{C}$. Verify both forms of DeMorgan's laws.

5

The Human–Machine Interface

Have you ever noticed that some machines are easier to use than others? Do you find that some products appeal to your sense of touch and sight, while others simply perform their designated function? Do some software programs seem extremely user friendly, while others seem impossible to operate? Easy-to-use items have one thing in common: an excellent human–machine interface. The human–machine interface defines the way in which a person interacts with an engineered product. It primarily involves the three senses of touch, sight, and hearing, and on rare occasions, smell and taste as well. A good product "feels" right. It is easy to use and becomes an extension of the user's motor and cognitive functions. Its features were not simply included just because they *could* be. Rather, the product was developed from the start with the needs of the user in mind. The designer of a good product thinks about *how* the device will be used in addition to *what* the product must do. A good product is one in which the entire function and purpose of the device has entered into the design process. Experienced engineers know that the human–machine interface is the first thing that should be considered during the initial stages of product development. Throughout this chapter, the word machine is defined to mean any

OBJECTIVES

In this chapter, you will:

- Learn how people interact with machines.
- Understand the importance of the human/machine interface.
- Examine case studies of good and bad human/machine interfaces.

mechanical, electrical, industrial, biomedical, structural, or software entity designed by an engineer.

5.1 HOW PEOPLE INTERACT WITH MACHINES

The pages of engineering history are littered with tales of products that fulfilled a useful function, but were hopelessly inept at providing an adequate human–machine interface. One canonical example of this principle is the programmable VCR. In a television commercial that aired in the early 1990's, a satisfied user of a voice-activated VCR controller proudly proclaims, "Hey, even *I* can't program my VCR by hand, and I have a *Master's* degree!" Indeed, much humor has been derived from the notion that seemingly intelligent people are content to let their VCR clocks blink at 12:00 rather than tackle the intricate maze of its programming functions. The problem with most under-utilized VCR features is seldom the intelligence of the user, but rather that most of us simply do not have the time to master hard-to-learn features provided by the manufacturer. Many other simple consumer products suffer from this deficiency. Examples include various versions of the digital watch, the programmable CD player, cellular telephones, PC operating systems, word processors, fax machines, coffee makers, and microwave ovens. Even the latest models of automobiles are laden with features that elude the owner, who is usually intent on simply driving the car. Without a simple, easy-to-remember sequence of programming steps, the features of the most intricate and complex machines lay idle most of the time.

5.2 ERGONOMICS

The human–machine interface influences our attitude toward the most mundane of devices. Some doors seem to require less effort to push open. Some knife-and-fork sets are more appealing to use than others. A favorite chair feels more comfortable than all the rest. The best of these products were designed with careful consideration of *ergonomics:* the science of how the body interacts with machines. Ergonomics focuses on the size, weight, and placement of objects and control devices that interact with the human body. The average span of the human arm, the swing frequency of a free limb, the height of the eyes above a tabletop, and the spacing between fingertips are all examples of things that must be considered when designing the physical layout of a product. Consider the case of the driver's position in the automobile, a product that has been fine tuned for over ninety years. Automakers give careful consideration to the placement of the steering wheel, gearshift lever, brake and accelerator pedals, heat and air-conditioning controls, radio knobs, rear view mirrors, and windshield aperture. Cars are designed for the average human body while maximizing as much as possible the range of physical attributes that can be accommodated. By how much should the position and height of a seat be adjustable? Will adding the expense of a tiltable steering wheel gain more market share because drivers on the fringe of the ergonomic range will still find the car appealing to drive? This approach to vehicle design has served the driving public for most of the history of the automobile, although occasionally unforeseen problems occur with this approach. The concern over airbags that surfaced in the mid 1990's illustrates a design principle run amok. The airbag is a marvelous device that can be credited with saving countless lives. But like all devices in the automobile cockpit, it was designed with the average driver in mind. It inflates at chest level, as defined for a person of average height. Children and short adults can receive a lethal blow to the head should an airbag inflate unexpectedly. This unforeseen problem led to the revised recommendation that children be confined to the rear seats of all automobiles having passenger-side airbags. While this advice provides a temporary solution, the hazard remains a problem for short adult drivers and front-seat passengers.

The study of ergonomics has produced a body of anthropometric data that can be used in designing anything that involves the interaction between a human and a machine. Tables of statistics on arm length, arm span, waist height, shoulder height, joint location, limb bending angles, and turning radii can be found in any of a number of references in the literature, including those listed at the end of this chapter. These data can be used to choose the size of knobs, location of openings, spacing of push buttons, and location of display devices. One goal should be to avoid awkward positions or physical actions on the part of the user. The software developers for many the first software programs written under the Microsoft™ DOS operating system, for example, took great advantage of the close proximity of the function keys to the QWERTY letter keys on the original 88-key keyboard of the first IBM XT™-and AT™-model computers to provide easy-to-use and effortless keyboard command sequences. Conversely, the master reset sequence was specifically chosen to be a very hard-to-engage key combination. The simultaneous pressing of the keys CTRL-ALT-DEL requires an unbelievably awkward finger span and use of two hands, making it unlikely that the user will accidentally reboot the machine during normal typing operations.

The rules of ergonomics are not difficult to learn. Although some of the more esoteric guidelines are the province of experts, most involve simple common sense. Button controls should be kept within finger span unless they are intended for occasional use only. Knobs and valves should be kept within an arm span, and those linked to a common function should be grouped together. Visual information should be kept within line of sight. Display devices should be located so as not to require constant turning or bobbing of the head. Command sequences or operations should follow a logical order and should be easy to remember. Some of these design principals are so ingrained in our subconscious expectations that we notice them only when they are violated. If you sat down behind the wheel of a car and found the ignition key switch on the left side of the steering wheel, you'd note that something did not feel quite right about the car's layout. If you opened a new software program and found the FILE pull-down menu in the upper right-hand instead of upper left corner, you might think it strange. The principles of ergonomics share a unique symbiosis with the shape of our bodies, the dimensions of our limbs and extremities, our shared expectations, and even elements of our culture.

Exercises 5.1

E1. Measure the shoulder-to-finger-tip arm span of your own body. Compare to the height of your desk and distance from its edge to your computer keyboard. Is there a correlation?

E2. The standard desk height is 76 cm (30 in). Measure the height of your waist above the floor. Is there any correlation?

E3. Measure your own floor-to-shoulder height and compare to the standard 106 cm (42 in) height of an electrical switch above the floor.

E4. Measure your own head diameter and compare to the unseparated span of the earphones of a stereo headset. Is there any correlation?

5.3 COGNITION

Nearly every product or engineered device requires learning on the part of the user before it can be operated. The typical user wants to have control over a product and fully understand its use. *Cognition* refers to the way in which the user learns about the device and masters its features, operation, and various characteristics. A well-engineered device will provide the user with a short learning curve and a consistent set of rules of operation. As an example of this principle, consider the now standard graphical user interface used in most computer software systems (e.g,, Windows™ or MacIntosh programs). Have you noticed that software programs always seem to incorporate pull-down menus? They appear because the user *expects* them to be there. A user will consider a program easy to use when it builds upon features learned from prior use of similar products. Another example can be found in the common automobile cited earlier. The gearshift lever is always in the same place, as are the lever to engage the directional signals and the button to sound the horn. Every car driver learns to operate these controls and expects them to be in the same place in every automobile. One aspect of car operation that has no consistency is the location and operation of the headlight switch. How often have you driven a strange car only to fumble momentarily while trying to figure out how to turn on the headlights or high beams?

In designing a product, device, or system, care should be given to making its operation easy to learn. As you approach the task of designing a new device, draw upon the operating principles of similar devices. Borrow functional sequences, structural details, or command sequences. Place controls where they are likely to be found on similar machines, or, at the very least, in logical places. We expect a light switch to be located along a room's interior wall, just inside the unhinged edge of the door. This location is logical, given the way one enters a room, and finding a switch placed anywhere else contradicts our learned behavior. The same can be said for the rotational direction of the volume control on radio and television sets. We subconsciously *expect* the volume to be increased by turning the knob clockwise. There is no hard logic to this choice, but it resonates with our learned notion that the rotational direction of the analog clock face corresponds to marching forward, or increasing something. Acknowledging the importance of cognition in engineering design requires that we design products and systems whose operation is well organized, easy to learn, easy to remember, and consistent in operation with other, similar products.

5.4 EXAMPLES OF GOOD AND BAD HUMAN–MACHINE INTERFACES

Engineers can learn a great deal from studying the successes and failures of other designers. While grand, large-scale failures such as falling bridges have gained much notoriety in educational circles (see Chapter 7, for example), small-scale failures are equally worthy of study, because they illustrate the impact of the small design decisions made by engineers every day. This issue is extremely important in the design of the human–machine interface. In this section, we cite several case studies of both well-designed and poorly designed human–machine interfaces. In the following discussions, the happy-face symbol (☺) will be used to denote an example of a good or improved human–machine interface, and the sad-face symbol (☹) will denote an example of a bad interface.

EXAMPLE 5.1: THE TELEPHONE DIAL ☹

Until the mid 1970s, almost all telephones had rotary dials. The integrated circuit technology needed to economically include DTMF Touch-Tone capability in all telephones was not yet available. A marvel of electromechanical design, the rotary dial, shown in Figure 5.1, allowed the user to insert a finger in the hole corresponding to the desired digit. Rotating the dial from the chosen digit to the hook and releasing it caused it to

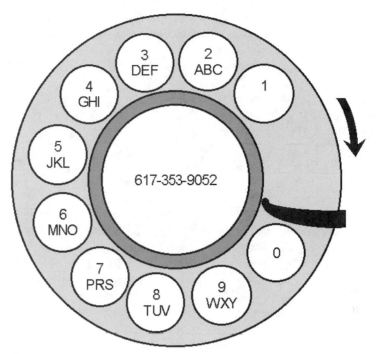

Figure 5.1. The rotary telephone dial.

send out the proper number of electrical current pulses to the central telephone station. The problem with the rotary dial was that it was mechanical, and therefore more prone to failure and much slower than its modern Touch-Tone replacement.

The Princess™ telephone, first marketed by Western Electric with a small rotary dial, placed the dialing mechanism, microphone, and earpiece all within the headset. This design represented a departure from the traditional desk phone that had been the standard since the 1940s. Its compact and utilitarian arrangement fulfilled the design goal of providing an instrument that did not require the user to return the headset to the base to make another call. This arrangement was ideal for someone lounging on a couch and making repeated phone calls. Might there have been a connection between the original name "Princess" and the image of a teenager in the 1960s? The engineers who conceived of the original Princess design could not have foreseen the pervasiveness of Touch-Tone dialing that exists today. Indeed, much of our modern world of communication, including modems, cellular telephones, cordless phones, voice mail, pagers, and telephone banking was made possible by the development of the Touch-Tone system. Nevertheless, the physical form of the Princess phone became the model for many of the Touch-Tone phones in use today. All cordless phones, cellular phones, and many table-top models are designed with the Touch-Tone keypad located in the center of the headset. The keypad-in-headset design, however, violates an important principle of ergonomic design. It makes the product much *more* difficult to use. In the typical scenario involving voice mail or automated telephone functions, the user first listens to an announcement, then presses the appropriate key selection. For most people, the latter operation requires looking at the keypad while pressing a key selection. Thus, the user must repeatedly hold the headset up to the ear, remove it from the ear and hold it in front of the face to press keys, and then return the headset to the ear to listen for the results or the next automated instruction. It is much easier to handle automated telephone dialing or voicemail instructions when the keypad is located in front of the user on a desktop base. (The latest technology circumvents the deficiencies of the keypad-in-headset phone by responding to spoken numerals, but these systems are not yet in widespread use.)

The first personal computer to gain widespread acceptance in the mid 1980's was the Intel-8086-processor-based IBM PC™. This early pioneer was soon replaced by the IBM XT, which included a hard disk drive, and later the IBM AT, which upgraded the processor to an Intel 80286. At this point in PC history, the mouse and the graphical user interface were in their developmental infancy and were not in widespread use.

As illustrated in Figure 5.2, the first XT and AT 88-key keyboards included ten function keys F1 through F10 located in two rows to the left of the standard QWERTY set. The control key was located immediately to the left of the letter A, the ALT key just below the left SHIFT key, and the caps-lock key just below the right SHIFT key. Early software developers took great advantage of this arrangement to write ergonomically pleasing user functions. For example, the authors of Wordstar™, one of the first word-processing programs, created cursor movement functions that could be initiated using the left hand only. By simultaneously pressing the CTRL key with the little ("pinky") finger of the left hand and one of the letters A, S, D, or F with another finger, all cursor movements could be initiated. Other commands involved combinations of left-hand QWERTY keys with one of the Fn function keys equally accessible to the left hand. These keyboard combinations were considered so efficient that they were adopted for other early software tools, including Borland's Turbo Basic™ and Turbo C™, as well as other software programs such as Visicalc™ and Lotus 1-2-3™. Even the first full-screen text editor included by Microsoft in its DOS operating system incorporated parts of the Wordstar key combinations into its command sequences.

Figure 5.2. The 88-key computer keyboard.

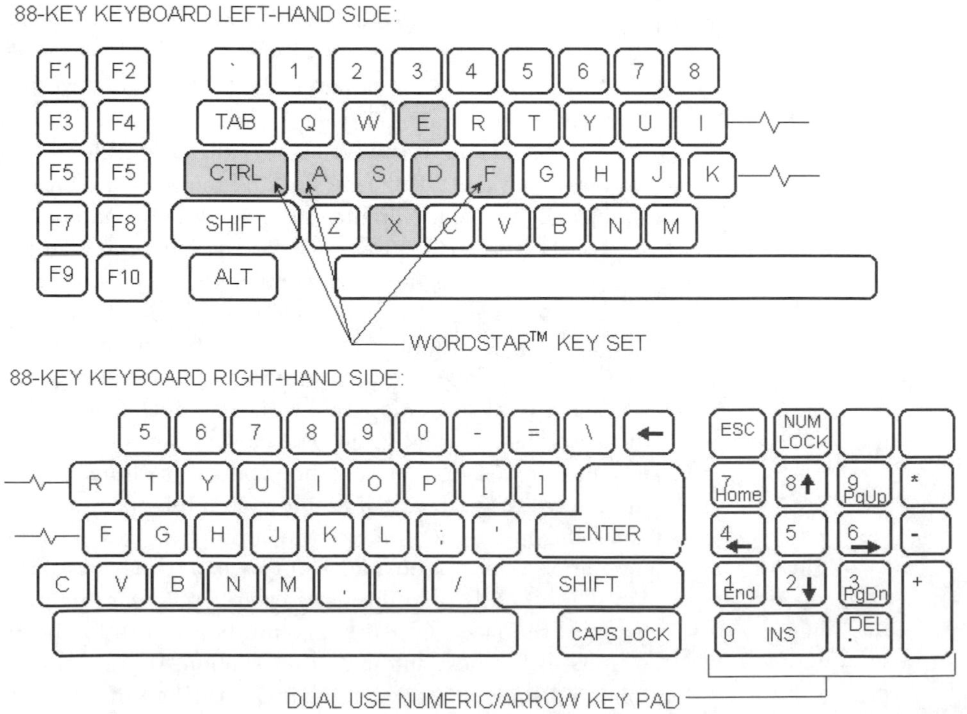

88-KEY KEYBOARD LEFT-HAND SIDE:

WORDSTAR™ KEY SET

88-KEY KEYBOARD RIGHT-HAND SIDE:

DUAL USE NUMERIC/ARROW KEY PAD

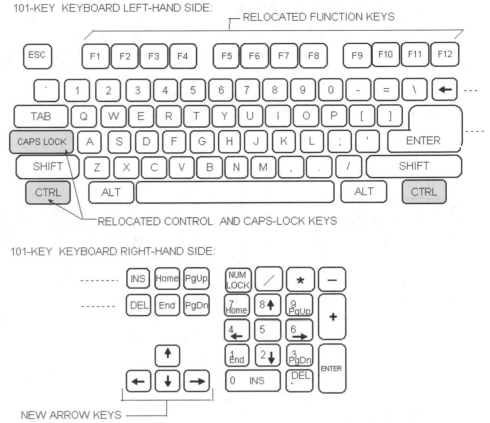

Figure 5.3. The redesigned 101-key computer keyboard.

The main problem with the 88-key keyboard was the dual use made of the numeric keypad located to the right of the QWERTY key set. These number keys also served as arrow, HOME, PgUP, PgDN, INS, and DEL keys. Engaging the numeric keypad required the user to press the NumLock key. As computers and software programs evolved, the new 101-key keyboard layout shown in Figure 5.3 soon became popular. The 101-keyboard quickly became the standard in the early 1990's. The dual-use numeric keypad persisted in the 101-key keyboard, but a new set of dedicated arrow, HOME, PgUP, PgDN, INS, and DEL keys appeared between the numeric keypad and the QWERTY set. This change considerably elongated the physical size of the keyboard, so to save space, the designers of the 101-key keyboard moved its F1 through F10 keys, along with two new F11 and F12 keys, to a row *above* the standard QWERTY set. This change in layout placed the Fn keys out of easy reach of the left hand, eliminating the ease with which entire sets of software commands could be accessed. Another change in the 101-key keyboard was the exchange of the CapsLock and CTRL keys. The reason for this change is unclear, but it moved the CapsLock key to the prime location adjacent to the letter A. This seemingly innocent change eliminated easy access to the CTRL key by the left pinky and signaled the demise of the very efficient Wordstar-like cursor control sequences. These original mainstays of keyboarding are seldom found in software written later than the early 1990's. They have been replaced with cumbersome Ins-Del and CTRL-Shift-Ins types of key sequences that require more complex finger contortions on the part of typists. Ease of function and ergonomic efficiency were displaced by progress in the name of added keys placed outside the finger span of the typical human hand.

Exercises 5.4

E5. Consider two telephones, one with the keypad in the headset, and one with the keypad on its desktop base. Estimate the extra time per call required when using the former to enter the numbers needed to access an automated flight arrival system for a major airline.

E6. Estimate the number of times that function keys are used when typing a typical paragraph on your word processor.

EXAMPLE 5.3:
THE
MARINER'S
COMPASS ☺

Good ergonomic design need not be limited to the world of high technology. The simple mariner's compass, used by sailors for centuries, provides an example of bad ergonomic design turned good. The first compasses were made by floating pieces of lodestone, a naturally occurring magnetized rock, in water or other liquids. Free to turn in any direction, the lodestone unfailingly pointed in the north-south direction, providing an important navigational aide to sailors. As technology progressed, the hard-to-find lodestone was replaced by magnetized iron and steel, eventually evolving into the floating compass rose design shown in Figure 5.4(a). This basic form of the compass, with its horizontally floating disk and glass bubble top, persists in many ships to this day. The problem with this design is that the disk can only be viewed from the top, because the printed surface of the compass is horizontal. The helmsman must look away from the horizon to glance down at the compass, leading to fatigue on long ocean voyages. Adding a mirror inclined at 45° to allow viewing from a horizontal line of sight does not help, because it reverses the apparent rotational direction of the compass and confuses the helmsman. The solution to this problem is a simple one. A much-improved, redesigned compass is shown in Figure 5.4(b). It replaces the floating disk with a cylindrical shell mounted on a jeweled pivot. The compass markings are printed on the vertical edge of the cylinder, allowing the entire compass to be mounted on a vertical wall of the cockpit or bridge. Mounted in a proper location, it can be viewed with only a slight downturn of the helmsman's

Figure 5.4.   The mariner's compass. a) original horizontal compass design; b) revised vertical-wall compass design.

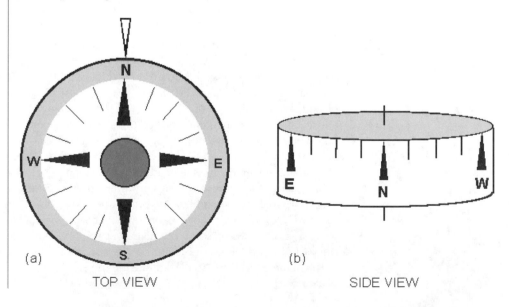

(a)

TOP VIEW

(b)

SIDE VIEW

eyes, rather than a complete lowering of the head. Fatigue during long voyages is greatly reduced.

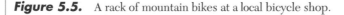

EXAMPLE 5.4:
THE
MOUNTAIN
BIKE ☺

The ubiquitous mountain bike, as shown in Figure 5.5, has become the choice of recreational bike riders everywhere. From the casual urban user to the most serious mountain trekker, riders are drawn to the mountain bike's comfort and ease of use. It's evolution to prominence provides an example of good ergonomic design. During the 1960's, the bicycle of choice by serious riders was a three-speed model that had a single gearshift lever mounted near the right-side grip of the handlebars. This bicycle, dubbed the "English bike" by Americans, replaced the balloon-tired "coaster brake" models, which were soon relegated to a kids-only status. A decade later, the ten-speed bicycle became the new rage of the bikeways. With its lightweight frame, large choice of gear ratios, and sleek tires, the ten speed was built for speed and efficiency. It soon became the favorite of teenagers, college students, and even adult riders. Its design was a verbatim offshoot of the bicycles that had been used for years by professional racers. As illustrated in Figure 5.6, the brake levers were mounted on the front of curved handlebars, and the gearshift levers were mounted on the foremost strut of the bicycle frame. The placement of these controls required the user to ride in a hunched, albeit curved and aerodynamic, position and to let go of the handlebars with one hand in order the change gears. Although the hunched-over position is very efficient for racing and long, cross-country riding, most recreational users found it uncomfortable and cumbersome. Small solutions, such as the placing of the gearshift levers on the handlebar post and the addition of brake-lever extensions that permitted the rider to hold the handlebars along the

Figure 5.5. A rack of mountain bikes at a local bicycle shop.

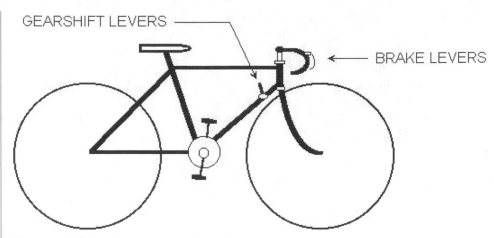

GEARSHIFT LEVERS

BRAKE LEVERS

Figure 5.6.   The classic ten-speed bicycle.

horizonal portion of the bar, rather than on its curved underwings, became available in the later days of the ten-speed's heyday. These improvements created safety hazards, however, as some of the brake extenders broke during hard braking.

Sometime in the late 1980's to early 1990's, the mountain bicycle began to appear in bicycle shops everywhere. It had a straight handlebar, its brake levers were within easy reach of an upright rider, and its gearshift levers were integrated into the hand grips or placed on easy-to-reach fingertip and thumb controls on the handlebars. The rider no longer had to let go of the handgrips to shift gears! By the mid 1990's, the mountain bike had become *the* bike of choice by all but serious cross-country riders and racers. Its strong frame and tires, designed to assault mountain paths, proved to be worthy of the bumpiest of city streets and came as an added bonus. But the success of mountain bicycles largely can be attributed to their designers' careful attention to ergonomics and their willingness to consider the needs, desires, and body shape of the end users.

EXAMPLE 5.5: THE TOGGLE LIGHT SWITCH ☺

Electricity first came into widespread use at the turn of the 20th century. Early methods of wiring and insulating were primitive, because the plastic materials that prevail today were not readily available. Indeed, most of them had not yet been invented. The first wall switches for lighting were large, rotational devices made from metal and ceramic. Sturdy, hearty, and virtually indestructible, these early pioneers of electrical switching had one major drawback. It was impossible to tell if the switch was on or off simply by looking at it. Later designs added a pointer to the rotary knob, but determining the position of the switch still required a careful gaze and some concentration. The much-improved toggle switch came into use sometime in the late 1930's to early 1940's. Its simple up-down design has persisted to this day, because its cognitive function has become second nature to us all: Up means on and down means off. (The only exception to this rule occurs in the three-way switches used for hallway lighting.) If you've ever encountered a room light switch that is mounted upside down, you've probably experienced some momentary sense that something was wrong as you tried to flip on the light.

The toggle motion of the modern-day light switch provides another important ergonomic benefit. Unlike its rotary predecessor, it can be switched with no hands at all. An elbow, knee, hip, stick, or even well-placed nose can do the job. This feature is helpful to users carrying shopping bags, individuals with special needs, or small children.

EXAMPLE 5.6:
THE
MISPLACED
TAPE
CASSETTE
SLOT ☹

A particular brand of radio/cassette player is marketed specifically for use in automobiles. The unit carries the four-letter logo of a well-known manufacturer of electronics associated with quality, hence many assume without question that it has been designed sensibly. Although the unit works well *electronically,* it has a serious ergonomic flaw. The slot for cassette tapes is located *above* the digital readout for the clock and radio station selection. The radio will not work with a tape inserted all the way into its slot, as one would do to play a tape, but a tape left partially out of the slot in the ejected position blocks the digital display from the driver's line of sight, as illustrated in Figure 5.7. This misplacement of the cassette slot means that a finished tape has to be removed entirely from the unit in order to view the clock or choose a radio station for listening, adding a needless distraction for the driver. The poor choice of functional layout suggests that the unit has been designed by desk-bound engineers who have never tried out the unit under actual driving conditions.

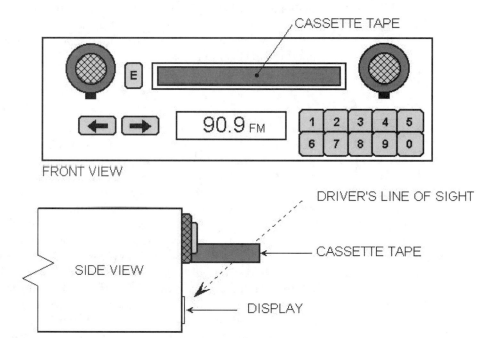

Figure 5.7.   The misplaced tape cassette slot. A tape cassette left in its slot in the ejected position blocks the digital readout from the driver's line of sight.

EXAMPLE 5.7:
THE DIGITAL
CLOCK ☹

Digital clocks can be found everywhere in our society. Integrated-circuit and display technologies have progressed to the point where economical electronic time pieces are generally far cheaper than their mechanical counterparts. The earliest models of digital clock incorporated power-hungry red LED displays into ac-powered desktop alarm clocks. Advances in integrated-circuit technology produced red LED watches that required the user to press a button to see the time. (Red LEDs require so much power that constant illumination via battery power is not feasible.) Soon thereafter, power-miserly liquid crystal displays were developed along with single-chip clock circuits, opening the door to cheap, reliable wrist watches and other timing devices. The attractiveness of electronic timekeeping technology and its robustness has caused digital time to all but replace the mechanical timekeeping devices that had served us well for over five hundred years.

Engineers who study and design by cognitive principles have come to realize that digital clocks have a fundamental disadvantage when compared with their analog counterparts. Although a digital clock will allow you to instantly know the exact time, most people are more interested in how much time is *left* before an event occurs. Examples include, "How much time is left before my class is over?" or "How many minutes do I have left to get to my appointment?" Viewing time from a digital clock requires that you do the necessary arithmetic in your head. This mental exercise can take a few extra seconds when time is provided to the nearest minute, for example, subtracting 9:48 from 10:00 to determine that you have but another twelve minutes to arrive at your ten o'clock destination. While your brain can readily do serial math, it functions much better as an image processor. In this task, the human brain is incredibly fast. No machine has yet been invented that can beat the human brain at general pattern recognition. Viewing the time 9:48 on analog clock hands allows you to estimate and subtract all in one glance as you instantly determine that you have about ten minutes left to get to your appointment. The contrast between digital and analog clocks provides an example of the often-ignored discord between technical progress and human cognition.

EXAMPLE 5.8: THE GYMNASIUM LIGHTING SYSTEM ☹

A large suburban high school outside of Boston, Massachusetts renovated its gymnasium. The school community was very proud of the upgraded facility, which included dual-use seating so that the room could be used for assemblies as well as for sports functions. As part of an overall energy-saving strategy at the school, the new gym included motion detectors on its overhead lights. The absence of motion on the central gymnasium floor caused the lights to gradually dim and then turn off.

The new gym was a huge success. It became the site of numerous basketball games, physical education classes, and even one school dance. The first time the room was used for a general assembly, however, a critical error on the part of the design engineers was discovered. A very famous and distinguished speaker had been invited to address the student body on an important social issue. The speaker's podium was set up at one end of the room, and students and faculty sat on the sideline bleachers and chairs. Ten minutes into his speech, the lights began to dim, eventually leaving the entire assembly in darkness. The motion detectors, sensing no movement in the center of the gym, proceeded to turn off the lights in the presence of the invited guest and the entire audience. The engineers had failed to consider all the ways in which the room would be used. They bothered only to focus on the technical issue of saving energy and had provided no manual overide for the motion detectors.

EXAMPLE 5.9: THE TRACKBALL ☹

Manufacturers of laptop computers are quick to point out the utility of the console-mounted trackball. The control device is always in the same place when needed. It requires no wire or hookup cable and is much more lightweight than a comparable mouse. In contrast, the standard desktop mouse is ideal from an ergonomic point of view. A swing of the hand over the span of the mousepad moves the cursor all the way across the monitor screen, and the buttons are placed directly under the fingers for instant access. Moving a cursor with a trackball is much less efficient. If the user sets the computer's mouse control to "fast," resolution is sacrificed because the entire screen span is confined to one-half turn of the trackball. If the control is set to "slow," resolution is better, but several spins of the trackball may be required to move the mouse to its intended location. This tradeoff makes use of the trackball tiring and tedious. The console mounted trackball provides a perfect example of a device that was marketed to address issues of packaging and portability without considering human ergonomics and its impact on efficiency and comfort.

EXAMPLE 5.10: THE DESIGNER LAVATORY FAUCET ☹

A famous maker of decorative plumbing fixtures has marketed the designer lavatory faucet shown in Figure 5.8. This fixture is a captivating composite of chrome and brass that adds to the decor of any well-appointed bathroom. The straight, cylindrical handles rise cleanly above the curved, inlaid base. This visually appealing faucet has but one problem: When the smooth, cylindrical handles become the slightest bit wet, they are nearly impossible to turn. The faucet handles provide a good example of form taking needless precedence over user function.

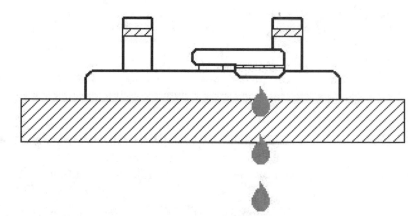

Figure 5.8.   The designer laboratory faucet.

EXAMPLE 5.11: THE SQUEEZE KETCHUP CONTAINER ☺

Many people enjoy eating at local, home-style restaurants. The atmosphere and friendliness in these owner-operated Mom-and-Pop establishments is always friendly and welcoming. One especially comforting item in these settings is the old-fashioned plastic squeeze ketchup container like the one shown in Figure 5.9. These little devices, with their just-right diameter and sharp pointed spouts, seem only to be found in home-style restaurants. They are rarely seen in supermarkets or national fast-food chains. While not glamorous or newsworthy, the lowly squeeze ketchup container is an example of superb engineering design. It's easy on the hand because its compliant walls are just the right thickness to make squeezing effortless. Its translucent walls are decorative, but they let the user see just how much ketchup is left inside. Its little pointed spout always dispenses the ketchup exactly where it's needed. Its glass bottle counterpart oft requires that one pour ketchup by madly thumping on the base of the overturned bottle. If a Nobel prize in ergonomics existed, it ought to go to whoever invented the plastic squeeze ketchup container.

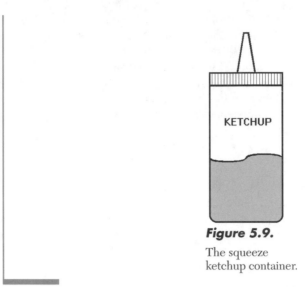

Figure 5.9.

The squeeze
ketchup container.

EXAMPLE 5.12:
THE CEREAL
DISPENSER ☹

A particular brand of plastic kitchenware offers a large container for storing breakfast cereal and other dry goods. The body of the container is much larger than the span of a typical hand grip, but the item is well designed and includes a recessed edge that is easily gripped for pouring out its contents. Its top includes a sealable flap that permits easy opening without requiring that the tight-fitting lid be removed entirely. The profile of the container as seen from the top, including the lid, is shown in Figure 5.10.

The design of this container has one major ergonomic flaw. The user can fill the container and put on the lid in either direction. No mechanism exists for forcing the user to put the opening flap on the side opposite the hand grip. If the lid is put on by an unfamiliar user, its flap will be placed on the wrong side of the container about 50 percent of the time, requiring the user to lift the container by its wide and awkward side.

Figure 5.10.   The cereal dispenser. Right Top: Lid flap placed on the correct side opposite the handle. Right Bottom: Lip flap placed incorrectly on the same side as the handle.

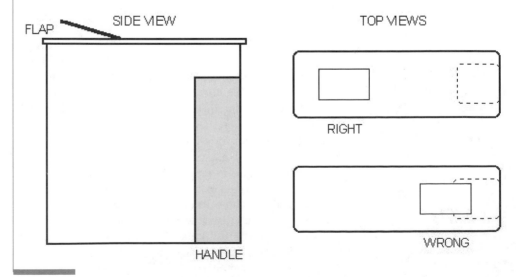

Exercises 5.4

E7. Consider the tape cassette shown in Figure 5.7. Draw a redesigned unit that would eliminate the line of sight problem described in Example 5.6.

E8. Consider the lavatory faucet depicted in Figure 5.8. Draw a revised design that would be easier to use but would retain the modern lines of the unit shown.

E9. Consider the cereal dispenser of Figure 5.10. Draw a modification to the product that would ensure that the cover is put on correctly each time.

EXAMPLE 5.13: THE HOSPITAL REVOLVING DOOR ☺

A large hospital recently renovated its main lobby. The previous building design had included revolving doors as a way to save energy by limiting air exchange while permitting rapid pedestrian traffic flow. Entering or leaving the hospital by wheelchair required use of large swinging doors located beside the bank of revolving doors. The problem with this arrangement is that *many* wheelchairs passed through the hospital every day. The hospital had (and still has) a policy of requiring that all released patients be delivered to their cars by wheelchair. In addition, many patients coming in for treatment were wheelchair-bound, and many parents wheeling baby strollers were forced to use the swinging doors as well.

The very clever design of Figure 5.11 solved the problem. By designing the revolving-door cavity to be an elongated shape with semicircular ends, and by designing an articulated door with outer wings, the architects and designers were able to produce a revolving door that accommodates wheelchairs and baby strollers along with walking pedestrian traffic. The new design expedites traffic flow and saves energy at the same time.

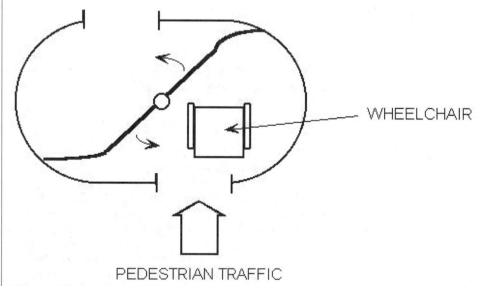

Figure 5.11. The hospital revolving door.

EXAMPLE 5.14: THE PAPER CLIP ☺

The simple office paper clip is a marvel of good ergonomic design. First patented around 1900, the basic paper clip has survived virtually unchanged for over 100 years. The different lengths of its long and sort tongues makes it extremely easy to clip over several pages using only one hand, and the different widths of its two tongues—one fits inside the other—gives it added clasping strength. Although manufacturers have attempted to come up with improved designs for the paper clip, none has succeeded in displacing the popularity of this indespensible office supply item.

PROFESSIONAL SUCCESS: BECOME AWARE OF THE HUMAN/MACHINE INTERFACE

Formal study can help you learn about the human/ machine interface, but direct observation in your own life is equally valuable. Be observant. If something seems difficult to use, try to figure out why. Think about how you might redesign the product to make it easier to use. Devise, in your own mind, modifications that will improve the product. By becoming aware of the machines and technology around you and identifying the design flaws of others, you'll become more skilled at designing your own human/machine interfaces.

KEY TERMS

Ergonomics   Cognition   Human/machine interface

REFERENCES

W. E. Woodson, B. Tillman, and P. Tillman, *Human Factors Design Handbook: Information and Guidelines for the Design of Systems, Facilities, Equipment, and Products for Human Use,* 2d ed. New York: McGraw-Hill, 1992.

K. R. Fowler, *Electronic Instrument Design: Architecting for the Life Cycle.* New York: Oxford University Press, 1996.

R. W. Bailey, *Human Performance Engineering: A Guide for System Designers,* Englewood Cliffs, NJ: Prentice-Hall, 1982.

Problems

1. Compile a list of five objects or devices that illustrate good ergonomic design. For every entry on your list, come up with a counterexample of poor ergonomic design.

2. Prepare a case study of an example of good or improved ergonomic design.

3. Prepare a case study of an example of poor ergonomic design.

4. Perform a telephone survey of fifty adults at random. Of those who own and use a VCR, ask how many regularly make use of its programmable features.

5. Perform a survey of people who use computers. Divide the list into those who use a conventional mouse, those who use a trackball, and those who use another pointing device. Of those in the first two categories, determine how many are happy with their pointing device and how many wish there existed something "better."

6. Design an experiment in which you draw a facsimile of the view as seen from the driver's seat of an automobile. Change the location of one feature of the layout (e.g., the location of the ignition key or gear shift lever). Show your drawing to a collection of test subjects. Record the amount of time that it takes each test subject to identify what is out of place.

7. Draw a diagram of a room with doors, windows, and furniture. Place the door handle on the same side as the hinges. Show your diagram to a number of test subjects. Record the amount of time that it takes each test subject to identify what is out of place.

Ergonomic Measurements

8. Measure a typical classroom chair in your school. Record the following dimensions: front-seat lip to floor; front-to-rear length of seat surface; front-seat edge to backrest; rear-seat edge to middle of backrest. Compare these dimensions to their corresponding body measurements on ten individuals and compile your data.

9. Find light switches in twenty different buildings (not just different rooms in the same building). Measure the height of the switch above the floor, and find the average value and the standard deviation. Now measure the elbow-to-ground height of thirty different people. Also determine the average value and the standard deviation, and compare to the light switch measurements.

10. Measure the shoulder-to-fingertip arm span of twenty individual adults. Determine the average length, the minimum, the maximum, and the standard deviation. Do twenty people provide a large enough sample? Should you obtain measurements of more people?

11. Measure the floor-to-eye height of ten seated people who work regularly at a computer. Compare their measured height to that of the center of the monitor screen on their computer. Ask each individual to estimate how long he or she can work at the computer before needing a break. See if there is any correlation between fatigue and monitor placement.

12. Find twenty five or more volunteers who are willing to walk a fixed distance of approximately 30 m (about 100 ft.). Count the total number of paces that each person requires to span the distance. Determine the average value and the standard deviation.

13. Measure the head circumference of thirty or more individuals. Determine the average, minimum, maximum, and most frequent head circumference measured to the nearest half centimeter.

Histograms

The following set of problems involves the use of the *bin histogram*. A histogram is a graphical plot that shows the number of members of an ensemble of data in various categories. For example, the histogram of Figure 5.12 shows the total number of occurrences of each letter of the alphabet in the text of this problem up to and including the end of this sentence. Similarly, the histogram of Figure 5.13 shows the distribution of end-of-semester grades in a particular engineering class. Engineers often use histograms to display data and sort information.

14. Ask one hundred full-grown people to tell you their height. Create a bin histogram that shows the distribution of heights in your sample pool.

Figure 5.12. Bin histogram showing the frequency of the letters of the alphabet appearing in paragraph text.

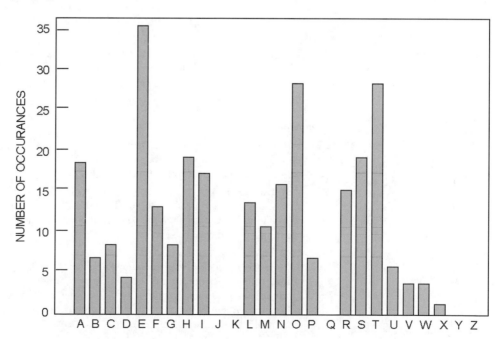

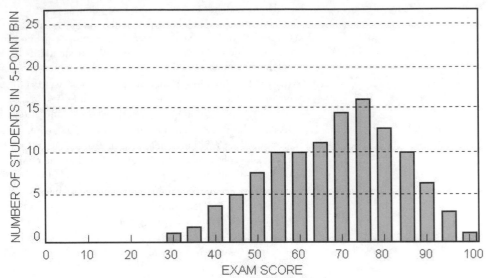

Figure 5.13. Bin histogram showing the distribution of grades in an engineering class.

15. Ask one hundred adults to tell you their weight. Create a bin histogram that shows the distribution of weights in your sample pool.

16. Measure the exact height and width of thirty different desks. Plot your data in bin histogram form.

17. Measure the length and width of at least fifty marked parking spaces in your community. Exclude handicapped parking spaces. Create two histograms, one for the length and one for the width, that show the distribution of parking space sizes.

18. Write a computer program to determine the keystroke frequency of each of the letters A–Z by someone typing The Gettysburg Address (or any similar document of your choice). Plot your results using a bin histogram.

19. Find twenty five people who ride bicycles. Measure the length of their legs from hip joint to the bottom of the foot with shoes on. Now measure the height of their bicycle seats above the ground. Plot a histogram of the ratio of seat height to leg length. Is there any obvious pattern? Is there any gender correlation?

20. The objective of this problem is to determine the most common choices of color for personal passenger cars. Find a location along a highway or busy road. Record the color of at least one hundred passing cars. Display your data, from most to least popular color, in bin histogram form.

Reaction Time

21. Write a computer program to track the keystrokes of someone typing a document of your choice into the computer. Find and record the average time that it takes the typist to activate each key after the entry of the preceding letter.

22. Perform the following test on several classmates. Prepare a card of simulated pushbuttons with the printed commands ON, OFF, LEFT, RIGHT, UP, DOWN, TURN LEFT, TURN RIGHT, and STOP. Prepare a second card in which the location and size of the buttons are the same, but the printed words have been replaced with the visual symbols shown in Figure 5.14.

    Now devise a set of keystroke sequences that simulate the navigation of a remote-control robot around a fictitious maze. Ask ten or more friends to press the simulated buttons upon your verbal commands. Keep track of how much time it takes each person to complete the sequence. Which do you think will lead to a faster reaction time: the printed keys or the graphically labeled keys?

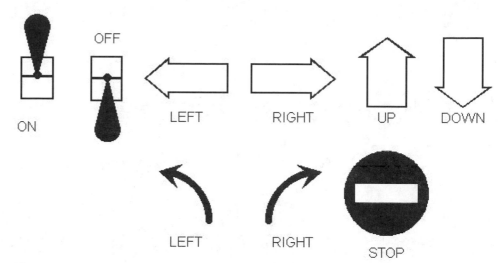

Figure 5.14. Visual commands for up, down, etc.

23. As you dictate the following passage, compare the time required for each of twenty people to write it on paper in longhand against the time required to type it into a computer: *Six saws saw six cypresses.* Translate this sentence into French and it comes out sounding like, *"See see see see see prey."*

24. Write a game on your computer that displays time from a made-up clock. For some participants, the time should be shown in analog form with hands. For others, it should be shown in digital form. As you repeatedly flash times on the screen, ask participants to press a key to stop the flashing when the displayed time is twelve (or whatever) minutes before another time that you've specified. For example, if you specified five minutes before 11:20, the correct answer would be 11:15. Have your computer program keep track of how much time elapses between the display of the correct answer and the pressing of the stop button. On average, is there a difference between reaction times to analog versus digital displays?

6

The Role of Failure in Engineering Design

Consider the following scenario that describes the experience of two students working on the Peak-Performance Design Competition introduced in Chapter 2:

> "The students had been working on their vehicle for almost a week. Having decided to enter the Peak-Performance Design Competition, they directed their design approach toward an articulated arm that would pick up the opposing vehicle and throw it off the track. They first outlined their design on paper and then tested it out on a computer-aided design (CAD) package available at one of their parent's place of work. They built all the parts in the school machine shop, using their CAD drawings as a guide. A sketch of their design is shown in Figure 6.1. The students had just finished putting together all fifty eight machined pieces and were delighted to find that the arm worked perfectly on the very first try!"

6.1 HOW ENGINEERS LEARN FROM MISTAKES

Wouldn't it be nice if engineering were so simple and foolproof as the scenario depicted in the preceding, fictitious paragraph? In the real world, almost nothing works correctly the first time, and getting things to work perfectly almost always takes longer than expected. Fabricated parts

OBJECTIVES

In this chapter, you will:

- Examine the role of failure in engineering design.
- Discuss classic design failures as case studies.
- Learn how to accept and utilize failure as part of the path to design success.

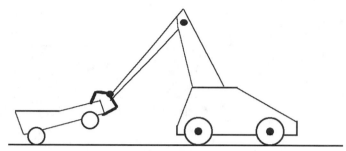

Figure 6.1. Hypothetical, impossible-to-build car design includes an articulated arm.

do not fit together, circuits have wiring errors, software modules have incompatibilities, and structural elements are incorrectly sized. Experienced engineers know that designs seldom work the first time and are never discouraged by initial failure.

Engineering tasks take longer than expected, because *failure* is an inevitable part of the design process. It's unreasonable to expect a new design to work the first time. When a device does not work as planned, it's a sure sign that something important has been overlooked. Perhaps two moving parts hit each other unexpectedly. Perhaps an electronic circuit doesn't work because stray effects were not included in the design model. A software program may fail because an unforeseen set of keystrokes leads to a logical dead end. A scale model of a bridge may reveal an overstressed support beam because a support pillar was omitted. A biomedical implant may be rejected because a tissue interaction was underestimated. Whatever the failure mode, it's better for a defect to appear *during* the design process than after it, when the device is in the field. Redesign after failure provides an important path to optimizing the final product and gives engineers needed time to correct deficiencies and work out bugs.*

A more realistic version of the paragraph that appeared at the start of this chapter might resemble the following:

"The students had been working on their Peak-Performance Design Competition vehicle for almost a month. Their first version of the car included an articulated arm designed to pick up the opposing vehicle and throw it off the track. After much trial and error, they succeeded in building an arm capable of lifting objects. At first, the students attempted to power the arm using the spring force from a single mousetrap, but after much testing, the students discovered that they had ignored frictional effects. Two mousetraps were required for adequate mechanical power. Eventually, they learned that powering the arm from a small electric motor and gear set provided much finer control of its movements. Their first attempt at assembly required that they return to the machine shop to redrill several holes they had put in the wrong place. Although the final version of the arm worked well on paper and also on a computer-aided design (CAD) package available at one parent's place of work, the arm failed miserably when mounted on the car. When an opposing vehicle was lifted into the air, the center of gravity fell outside the maximum allowed wheelbase of the vehicle, causing *both*

---

*The word "bug" was coined by Adm. Grace Hopper in 1945. In the early days of computers, logic gates were made from electromechanical relay switches rather than transistors. A moth flew inside one such computer, got stuck between two relay contacts, and prevented seemingly closed relay contacts from making a true connection. The computer malfunctioned because of the "bug."

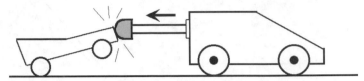

Figure 6.2. More realistic battering ram design as offensive strategy.

vehicles to topple over and fall off the track. The students had to abandon their arm design and eventually settled on the battering ram of Figure 6.2 for their offensive strategy."

6.2 CLASSIC DESIGN FAILURES: CASE STUDIES

The pages of engineering history are full of examples of design flaws that escaped detection in the design phase only to reveal themselves once the device was in actual use. Although all engineers are plagued by hidden design flaws from time to time, a few failure cases have become notorious because they affected many people, caused great property damage, or led to sweeping changes in engineering practice. In this section, we review several design failures from the annals of engineering lore. Each event involved the loss of human life or major destruction of property, and each was caused by engineering design failures. The mistakes were made by engineers who did the best they could, but had little prior experience or had major lapses in engineering judgement. After each incident, similar disasters were averted, because engineers were able to study the *causes* of the problems and establish new or revised engineering standards and guidelines. Studying these classic failures and the mistakes of the engineers who caused them will help you to avoid making similar mistakes in your own work.

The failure examples to follow all had dire consequences. Each occurred once the product was in use, long after the initial design, test, and evaluation phases. It's always better for problems to show up *before* the product has gone to market. Design problems can be corrected easily during testing, burn-in, and system evaluation. If a design flaw shows up in a product or system that has already been delivered for use, the consequences are far more serious. As you read the examples of this section, you might conclude that the causes of these failures in the field should have been obvious, and that failure to avoid them was the result of some engineer's carelessness. Indeed, it's relatively easy to play "Monday-morning quarterback" and analyze the cause of a failure *after* it has occurred. But as any experienced engineer will tell you, spotting a hidden flaw during the test phase is not always easy when a device or system is complex and has many parts or subsystems that interact in complicated ways. Even simple devices can be prone to hidden design flaws that elude the test and evaluation stages. Indeed, one of the marks of a good engineer is the ability to ferret out flaws and errors *before* the product finds its way to the end user. You can help to strengthen your abilities with the important intuitive skill of flaw detection by becoming familiar with the classic failure incidents discussed in this section. If you are interested in learning more details about any of the case studies, you might consult one of the references listed at the end of the chapter.

Case 1: Tacoma Narrows Bridge

The Tacoma Narrows Bridge, built across Puget Sound in Tacoma, Washington in 1940, was the longest suspension bridge of its day. The design engineers copied the structure of smaller, existing suspension bridges and simply built a longer one. As had been done with countless shorter spans, support trusses deep in the structure of the bridge's framework were omitted to make it more graceful and visually appealing. No calculations were done to prove the structural integrity of a longer bridge lacking internal support trusses. Because the tried-and-true design methods used on shorter spans had been well tested, the engineers assumed that these design methods would work on longer spans. On November 7, 1940, during a particularly windy day, the bridge started to undulate and twist, entering into the magnificent torsional motion shown in Figure 6.3. After several hours, the bridge crumbled as if it were made from dry clay; not a piece remained between the two main center spans.

What went wrong? The engineers responsible for building the bridge had relied on calculations made for smaller bridges, even though the assumptions behind those calculations did not apply to the longer span of the Tacoma Narrows Bridge. Had the engineers heeded some basic scientific intuition, they would have realized that three-dimensional structures cannot be directly scaled upward without limits.

Case 2: Hartford Civic Center

The Hartford Civic Center was the first of its kind. At the time of its construction in the mid 1970's, no similar building had been built before. It's roof was made from a space frame structure of interconnected rods and ball sockets, much like a child's construction toy. Rods numbering in the hundreds were interconnected in a visually appealing geodesic pattern like the one shown in Figure 6.4. Instead of performing detailed hand calculations, the design engineers relied on the latest computer models to compute the loading on each individual member of the roof structure. Recall that computers in those days were much more primitive than those we enjoy today. The PC had not yet been invented, and all work was performed on large mainframe computers.

Figure 6.3.   The Tacoma Narrows Bridge in torsional vibration.

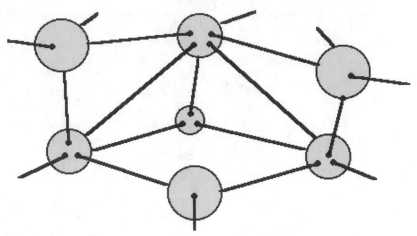

Figure 6.4.   Geodesic, rod-and-ball socket construction.

On January 18, 1978, just a few hours after the center had been filled to capacity with thousands of people watching a basketball game, the roof collapsed under a heavy snow load, demolishing the building. Miraculously, no one was hurt in the collapse.

Why did the collapse occur? Some attribute the failure to the engineers who designed the civic center and chose not to rely on their basic judgement and intuition gleaned from years of construction practice. Instead, they relied on computer models of their new space frame design. These computer models had been written by programmers, not structural engineers, during the days when computer modeling was in its infancy. The programmers based their code algorithms on structural formulas from textbooks. Not one of the programmers had ever actually built a roof truss. All failed to include basic derating factors at the structural joints to account for the slight changes in layout (e.g., minor variations in angles, lengths, and torsion) that occur when a complex structure is actually built. The design engineers trusted the output of computer models that never had been fully tested on actual construction. Under normal roof load, many ball-and-socket joints were stressed beyond their calculated limits. The addition of a heavy snow load to the roof load proved too much for the structure to bear.

Case 3: Space Shuttle Challenger

The NASA Space Shuttle Challenger blew up during launch on a cold day in January 1986 at Cape Kennedy (Canaveral) in Florida. Thousands witnessed the explosion as it happened. (See Figure 6.5). Hundreds of millions watched news tapes of the event for weeks afterwards. After months of investigation, NASA traced the problem to a set of O-rings used to seal sections of the multisegmented booster rockets. The seals were never designed to be operated in cold weather, and on that particular day, it was about 28°F (–2°C; a very cold day for Florida). The frozen O-rings were either too stiff to properly seal the sections of the booster rocket or became brittle and cracked due to the unusually cold temperatures. Flames spewed from an open seal during acceleration and ignited an adjacent fuel tank. The entire spacecraft blew up, killing all seven astronauts on board, including a high school teacher. It was the worst space disaster in U.S. history.

In using O-rings to seal adjacent cylindrical surfaces, such as those depicted in Figure 6.6, the engineers had relied on a standard design technique for rockets. The Challenger's booster rockets, however, were much larger than any on which O-rings had been used before. This factor, combined with the unusually cold temperature, brought the seal to its limit, and it failed.

Figure 6.5. The Space Shuttle Challenger explodes during launch. (*Photo courtesy of RJS Associates.*)

There was, however, another dimension to the failure. *Why* had the booster been built in multiple sections, requiring O-rings in the first place? The answer is complex, but the cause was largely attributable to one factor: The decision to built a multi-section booster was, in part, *political.* Had engineering common sense been the sole factor, the boosters would have been built in one piece without O-rings. Joints are notoriously weak spots, and a solid body is almost always stronger than a comparable

Figure 6.6. Schematic depiction of o-ring seals.

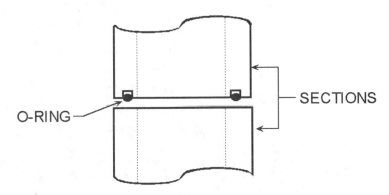

one assembled from sections. The manufacturing technology existed to build large, one-piece rockets of appropriate size. But a senator from Utah lobbied heavily to have the contract for constructing the booster rockets awarded to a company in his state. It was not physically possible to transport a large, one-piece booster rocket all the way from Utah to Florida over existing rail lines. Trucks were too small, and no ships were available that could sail to land-locked Utah, which lies in the middle of the United States. The decision by NASA to award the contract to the Utah company resulted in a multisection, O-ring-sealed booster rocket whose smaller pieces would easily be shipped by rail or truck.

Some say the catastrophe resulted from a lack of ethics on the part of the design engineers who suspected the O-ring design of having potential problems. Some say it was the fault of NASA for succumbing to political pressure from Congress, its ultimate funding source. Others say it was just an unusual convergence of circumstances, since neither the Utah senator nor the design engineers knowingly advocated for a substandard product. The sectioned booster had worked flawlessly on many previous shuttle flights that had not been launched in subfreezing temperatures. Still others say that by putting more weight on a political element of the project, rather than on pure engineering concerns, the engineers were forced into a less-than-desirable design concept that had never before been attempted on something so large.

Case 4: Kansas City Hyatt

If you've ever been inside a Hyatt hotel, you know that their internal architectures are very unique. The typical Hyatt hotel has cantilevered floors that form an inner trapezoidal atrium, and the walkways and halls are open, inviting structures. There's nothing quite like the inside of a Hyatt. In the case of the Kansas City Hyatt, first opened in 1981, the design included a two-layer, open-air walkway that spanned the entire lobby in midair, from one balcony to another. During a party that took place not long after the hotel opened, the walkway was filled with people dancing in time to the music. The weight and rhythm of the load of people, perhaps in resonance with the walkway, caused it to collapse suddenly. Over one hundred people died, and the event will be remembered forever in the history of hotel management. Although the hotel eventually reopened, to this day the walkway has never been rebuilt.

The collapse of the Hyatt walkway is a classic example of failure due to lack of construction experience. In this case, however, the error originated during the *design* phase, not the construction phase. In order to explain how the walkway collapsed, consider the sketch of the skeletal frame of the walkway, as specified by the design engineer, shown here in Figure 6.7.

Each box beam was to be held up by a separate nut threaded onto a suspended steel rod. The rated load for each nut-to-beam joint was intended to be above the maximum weight encountered during the time of the accident. What's wrong with this picture? The problem is that the structure as specified was not a realistic structure to build. The design called for the walkway's two decks to be hung from the ceiling by a single rod at each support point. The rods were made from smooth steel having no threads. Threading reduces the diameter of a rod, so it's impossible to get a nut to the middle of a rod unless the rod is threaded for at least half its length. In order to construct the walkway as specified, each rod would have to be threaded along about 20 feet of its length, and numerous of rods were needed for the long span of the walkway. Even with an electric threading machine, it would have taken days to thread all the needed rods. The contractor who actually built the walkway proposed a modification to the construction so that only the very ends of the rods would have to be threaded. The modification is illustrated in Figure 6.8.

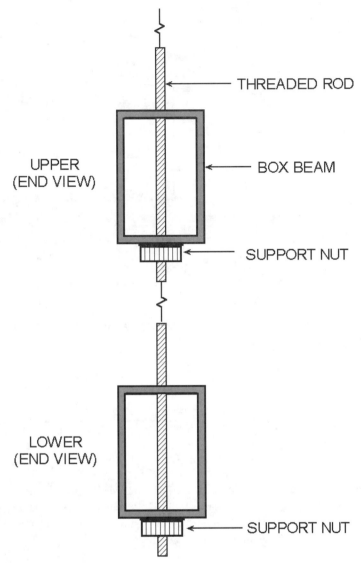

THREADED ROD

UPPER
(END VIEW)

BOX BEAM

SUPPORT NUT

LOWER
(END VIEW)

SUPPORT NUT

Figure 6.7. Kansas City Hyatt walkway support structure as designed.

The problem with this modification is that the nut (A) at the lower end of the upper rod now had to support the weight of *both* walkways. A good analogy would be two mountain climbers hanging onto a rope. If both grabbed the rope simultaneously, but independently, the rope could hold their weight. If the lower climber grabbed the ankles of the upper climber instead of the rope, however, the upper climber's hands would have to hold the weight of *two* climbers. Under the full, or maybe excessive, load conditions of that day, the weight on nut (A) of the Hyatt walkway was just too much, and the joint gave way. Once the joint on one rod failed, the complete collapse of the rest of the joints and the entire walkway quickly followed.

Some attributed the fatal flaw to the inexperience of the design engineer who specified single rods requiring 20 feet of threading. Others blamed it on lack of experience. Yet others blamed both the junior engineer, who signed off on the modifications presented by the construction crew at the construction site, and the senior engineer,

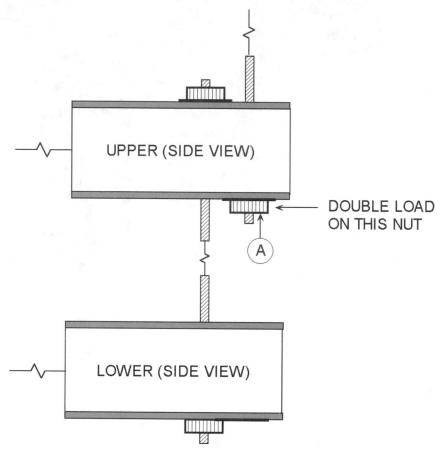

Figure 6.8.   Kansas City Hyatt walkway support structure as actually built.

who should have communicated to the junior engineer the critical nature of the rod structure as specified. Perhaps both engineers lacked seasoning—the process of getting their hands dirty on real construction problems as a way of gaining a feeling for how things are made in the real world.

Regardless of who was at fault, the design also left little room for *safety margins*. It's common practice in structural design to leave *at least* a factor-of-two safety margin between the calculated maximum load and the expected maximum load on a structure. The safety margin allows for inaccuracies in load calculations due to approximation, random variations in material strengths, and small errors in fabrication. Had the walkway included a safety margin of a factor of two or more, the doubly stressed joint on the walkway might not have collapsed, even given its modified construction. The design engineers specified a walkway structure that was possible, but not practical, to build. The construction supervisor, unaware of the structural implications, but wishing to see the job to completion, ordered a small, seemingly innocent, but ultimately fatal, change in the construction method. Had but one of the design engineers ever spent time working on a construction site, this shortcoming might have been discovered. Errors such as the one that occurred at the Kansas City Hyatt can be prevented by including workers from all phases of construction in the design process, ensuring adequate communication between all levels of employees, and adding far more than minimal safety margins where public safety is at risk.

Case 5: Three Mile Island

Three Mile Island was a large nuclear power plant in Pennsylvania (See Figure 6.9.) It was the sight of the worst nuclear accident in the United States and nearly comparable to the total meltdown at Chernobyl, Ukraine. Fortunately, the incident at Three Mile Island resulted in only a near miss at a meltdown, but it also led to the permanent shutting down and trashing of a billion-dollar electric power plant and significant loss of electrical generation capacity on the power grid in the eastern United States.

On the day of the accident, a pressure buildup occurred inside the reactor vessel. It was normal procedure to open a relief valve in such situations to reduce the pressure to safe levels. The valve in question was held closed by a spring and was opened by applying voltage to an electromagnetic actuator. The designer of the electrical control system had made one critical mistake. As suggested by the schematic diagram shown in Figure 6.10, indicator lights in the control room lit up when power was applied to or removed from the valve actuator coil, but the control panel gave no indication about the *actual* position of the valve. After a pressure-relief operation, the valve at Three Mile Island became stuck in the open position. Although the actuation voltage had been turned off and lights in the control room indicated the valve to be closed, it was actually stuck open. The mechanical spring responsible for closing the valve did not have enough force to overcome the sticking force. While the operators, believing the valve to be closed, tried to diagnose the problem, coolant leaked from the vessel for almost two hours. Had the operators known that the valve was open, they could have closed it manually or taken other corrective measures. In the panic that followed, however, the operators continually believed their control-panel indicator lights and thought that the valve was closed. Eventually the problem was contained, but not before a rupture nearly occurred in the vessel. Such an event would have resulted in a complete core meltdown and spewed radioactive gas into the atmosphere. Even so, damage to the reactor core was so severe that the plant was permanently shut down. It has never reopened.

The valve actuation system at Three Mile Island was designed with a poor human–machine interface. The ultimate test of such a system, of course, would be during an emergency when the need for absolutely accurate information would be critical. The operators assumed that the information they were receiving was accurate, while in reality it was not. The power plant's control panel provided the key information by inference, rather than by direct confirmation. A better design would have been one that

Figure 6.9.   Three Mile Island power plant.

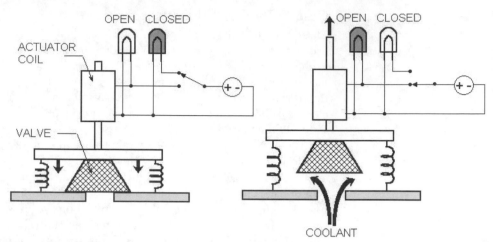

Figure 6.10. Valve indicator system as actually designed.

included an independent sensor that unambiguously verified the true position of the valve, as suggested by the diagram of Figure 6.11.

Case 6: USS Vincennes

The Vincennes was a U.S. missile cruiser stationed in the Persian Gulf during the Iran-Iraq war. On July 3, 1988, while patrolling the Persian Gulf, the Vincennes received two IFF (Identification: Friend or Foe) signals on its Aegis air-defense system. Aegis was the Navy's complex, billion-dollar, state-of-the-art information-processing system that displayed more information than any one operator could possibly hope to digest. Information saturation was commonplace among operators of the Aegis system. The Vincennes had received two IFF signals, one for a civilian plane and the other for a military plane. Under the pressure of anticipating a possible attack, the overstimulated operator misread the cluttered radar display and concluded that only one airplane was approaching the Vincennes. Repeated attempts to reach the nonexistent warplane by radio failed. The captain concluded that his ship was under attack and made the split-second decision to have the airplane shot down. Two hundred and ninety civilians died needlessly.

Figure 6.11. Valve indicator system as it should have been designed and built.

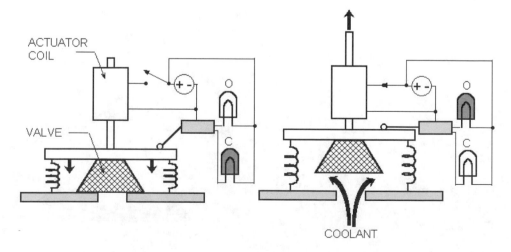

What caused this catastrophic outcome? Was it bad military judgement? Was it an operating error? Were the engineers who designed the system at fault? The Navy officially attributed the accident to "operator error" by an enlisted sailor, but in some circles the blame was placed on the engineers who had designed the system. Under the stress of possible attack and deluged with information, the operator simply could not cope with an ill-conceived human–machine interface designed by engineers. Critical information, being needed most during crisis situations, should have been uncluttered and easy to interpret. The complex display of the Aegis system was an example of something that was designed just because it was technically possible. It resulted in a human–machine interface that became a weak link in the system.

Case 7: Hubble Telescope

The Hubble is an orbiting telescope that was put into space at a cost of over a billion dollars. Unaffected by the distortion experienced by ground-based telescopes due to atmospheric turbulence, the Hubble has provided spectacular photos of space and has made possible numerous astronomical discoveries. Yet the Hubble telescope did not escape design flaws. Of the many problems that plagued the Hubble during its first few years, the most famous was its improperly fabricated mirrors. They were distorted and had to be corrected by the installation of an adaptive optic mirror that compensated for aberations. The repairs were carried out by a NASA Space Shuttle crew. Although this particular flaw is the one most often associated with the Hubble, it was attributed to sloppy mirror fabrication rather than to a design error. Another, less-well-known design error more closely illustrates the lessons of this chapter. The Hubble's solar panels were deployed in the environment of space, where they were subjected to alternate heating and cooling as the telescope moved in and out of the earth's shadow. The resulting expansion and contraction cycles caused the solar panels to flap like the wings of a bird. Attempts to compensate for the unexpected motion by the spacecraft's computer-controlled stabilizing program led to a positive feedback effect which only made the problem worse. Had the design engineers anticipated the environment in which the telescope was to be operated, they could have compensated for the heating and cooling cycles and avoided the problem. This example illustrates that it's difficult to anticipate all the conditions under which a device or system may be operated. Nevertheless, extremes in operating environment often are responsible for engineering failures. Engineers must compensate for this problem by testing and *retesting* devices under different temperatures, load conditions, operating environments, and weather conditions. Whenever possible (though obviously not possible in the case of the Hubble), a system should be developed and tested in as many different environmental conditions as possible if a chance exists that those conditions will be encountered in the field.

Case 8: De Haviland Comet

The De Haviland Comet was the first commercial passenger jet aircraft. A British design, the Comet enjoyed many months of trouble-free flying in the 1950's until several went down in unexplained crashes. Investigations of the wreckages suggested that the fuselages of these planes had ripped apart in mid flight. For years, the engineers assigned the task of determining the cause of the crashes were baffled. What, short of an explosion, could have caused the fuselage of an aircraft to blow apart in flight? No evidence of sabotage was found at any of the wreckage sites. After some time, the cause of the crashes was discovered. No one had foreseen the effects of the numerous pressurization and depressurization cycles that were an inevitable consequence of takeoffs and landings. Before jet aircraft, lower altitude airplanes were not routinely operated under pressure. Higher altitude jet travel brought with it the need to pressurize the cabin. In the case of the Comet, the locations of the rivets holding in the windows developed fatigue cracks, which, after many pressurization and depressurization cycles, grew into large, full-blown cracks in the fuselage. This mode of failure is depicted in Figure 6.12.

Had the design engineers thought about the environment under which the finished product would be used, the problem could have been avoided. Content instead with laboratory stress tests that did not mimic the actual pressurization and depressurization cycles, the engineers were lulled into a false sense of security about the soundness of their design. This example of failure again underscores an important engineering lesson: Always test a design under the most realistic conditions possible. Always assume that environmental conditions will affect performance and reliability.

Preparing for Failure in Your Own Design

First-time designs often betray previously hidden flaws after an initial period of successful use. Design flaws eventually show up because the operating environment changes, a previously untried sequence of events occurs, or weak points in the design encounter repeated stress. Sometimes, failure occurs just because of plain old statistics. As the saying goes, "If something is bound to fail, it *will* fail sooner or later." After a failure occurs, it's the engineer's job to determine the cause, fix what's wrong, and begin tests again. At the same time, it's up to the design engineer to identify as many of the bugs and weaknesses as possible during the early phases of the design process. Thorough testing and retesting under all sorts of operating conditions is essential. An unsuccessful first prototype presents an excellent opportunity to discover and weed out bugs before the final version of the device is put to market.

In the commercial sector, the rush to bring a product to market ahead of competitors puts pressure on the engineer to complete the test and evaluation phases as quickly as possible. For this reason, many consumer products, including automobiles, computers, and software, develop problems soon after they are released. If you purchase one of these items during the first year of issue, expect to find bugs and weaknesses that were not discovered on the factory floor.

Despite this admonition, you should be eager to apply your engineering design skills to new technology and innovation. If all engineers were content to stay with tried-and-true designs, technical progress would cease. Understanding when to stay within the

Figure 6.12. Stress cracks around the window rivets of the DeHaviland Comet.

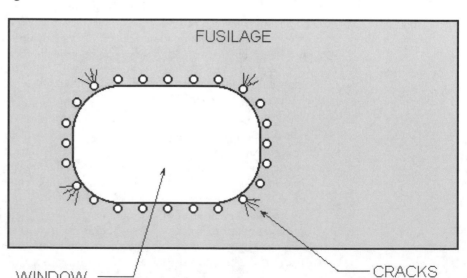

bounds of a traditional design and when to move on to new creative frontiers requires experience, practice, and intuition. When you encounter failure in your own design projects, do not be discouraged. Recognize failure as an inevitable part of the design process, and use it to learn, discover, and expand your foundation of engineering knowledge.

KEY TERMS

Failure   Flaw   Bugs
Safety margin

REFERENCES

J. FELD, KENNETH and L. Carper, *Construction Failure—Wiley Series of Practical Construction Guides,* New York: John Wiley & Sons, 1996.

E. S. FERGUSON, "How Engineers Lose Touch," *Invention and Technology.* Vol. 8 (3), Winter 1993, pp.16–24.

H. PETROSKI, *To Engineer is Human: The Role of Failure in Successful Design.* New York: Vintage Books, 1992.

H. PETROSKI, *Design Paradigms: Case Histories of Error and Judgement in Engineering.* Cambridge: Cambridge Univ. Press, 1994.

D. D. A. PIESOLD, *Civil Engineering Practice: Engineering Success by Analysis of Failure.* New York: McGraw-Hill, 1991.

R. UHL, G. M. DAVIDSON, and K. A. ESAKLUL, *Handbook of Case Histories in Failure Analysis.* ASM International, 1992.

P. VISWANADHAM and P. SINGH, *Failure Modes and Mechanisms in Electronic Packages.* Chapman & Hall, 1997.

Problems

1. Identify a product or system in your own experience that has failed. Write a short summary of the cause of the failure and how you might improve upon the design.

   Look up and write a synopsis of the following classic engineering failure incidents:

2. Exxon Bayway refinery, Linden, New Jersey (1990)
3. General Electric rotary compressor refrigerators (1990)
4. Green Bank radio telescope (1989)
5. Union Carbide chemical leak, Bhopal, India (1984)
6. Korean Airlines Flight 007 (1983)
7. Interstate 95 Bridge, Mianus River, Connecticut (1983)
8. Alexander L. Kielland oil platform, North Sea (1980)
9. American Airlines DC-10 (1979)
10. Skylab (1979)
11. The New York City Power Blackout (1976)
12. The windows in the John Hancock Tower, Boston, Massachusetts (1976)
13. Big Ben, London (1976)
14. Bay Area Rapid Transit (BART) (1973)
15. Point Pleasant Bridge, Ohio River, Ohio-West Virginia (1967)
16. The Apollo 1 capsule fire (1967)
17. The Great Northeast Power Blackout (1965)
18. Quebec City bridge (1907)
19. The Johnstown flood (1889)
20. Liberty Bell, Philadelphia (1835)

7

Effective Communication

Imagine that you have just purchased a new, top-of-the-line computer with the fastest available processor, many megabytes of memory, a huge disk drive, and a large, sophisticated sound card. When you get the computer home from the store, you notice that the model you bought has been shipped without a monitor. In fact, the computer is missing entirely a socket where a monitor can be connected. In all other respects, the computer is state-of-the-art and in perfect condition. What would your reaction be? You probably would take the computer back to the store claiming, of course, that any computer, no matter how fast or powerful, is useless without a means to extract the information it produces. Now imagine a similar scenario in which the computer has a monitor but comes with a very poor modem capable of a maximum data transfer rate of only 300 baud (about thirty printed characters per second). Such a slow modem would indeed limit the efficiency with which your computer could talk to other computers over the Internet. Reading electronic mail (e-mail) would be hopelessly slow, and accessing Web pages would be next to impossible. You probably would take this second computer back to the store also, claiming it to be a powerful machine that is incapable of communicating with others.

OBJECTIVES

In this chapter, you will learn about:

- Writing effective e-mail messages and memos.
- Preparing for formal and informal presentations.
- Writing long reports and journal papers.
- Preparing an instruction manual.
- Identifying the characteristics of good oral and written communication.
- Studying examples of good and bad writing styles.

People are a little bit like computers in this respect. The smartest person in the world, the fastest thinker, the most prolific scientist or engineer, is severely handicapped without an ability to communicate with others. The famous physicist Dr. Steven Hawking, author of *A Brief History of Time** and other works on cosmology, has been acknowledged as one of the most brilliant minds of modern physics. A debilitating disease known as ALS took away his ability to speak and move his limbs and facial muscles, cutting off all normal means of communication. His thoughts and ideas have come to the world one agonizing word at a time by way of synthesized computer speech. How much more the world would understand about the nature of the universe were Stephen Hawking able to communicate using normal human speech and body language

7.1 THE IMPORTANCE OF GOOD ORAL AND WRITTEN COMMUNICATION

Asked what the most important single skill is for new engineers in the workplace, most every employer will state emphatically: *communication skills!* The best software programmer in the world will do a poor job if the ideas behind the software are not well understood by the user. The most proficient mechanical designer will fail if the structure being designed is set up incorrectly or is used in an improper manner. Communication skills are so important that the Accreditation Board for Engineering and Technology, the national organization responsible for accrediting programs in schools of engineering in the U.S. and Canada, has listed oral and written communication as mandatory elements for all engineering programs regardless of discipline. Listening also is a very important skill for students of all types. Engineering departments have risen to the challenge by teaching these skills in many different ways, from writing workshops and courses taught by English departments to required oral presentations by students in key core courses.

In this chapter, we cover some of the most basic of oral and written communication skills. Although not all inclusive, the scenarios presented in this chapter cover many of the communication situations in which you might find yourself during your engineering career. Topics include preparing for meetings, conferences and presentations, drafting short memos, writing letters and electronic mail, and writing long technical reports and journal papers.

7.2 PREPARING FOR INFORMAL MEETINGS AND CONFERENCES

Whenever you get together to socialize, you are participating in an informal meeting. You also participate in an informal meeting when you meet with your boss or co-workers to discuss the status of your latest project. Are these two events similar? Should you approach both with the same level of preparation? In the first case, you'd feel silly if you prepared beforehand for a social gathering with your friends. The purpose of such a gathering would be to relax and dispense with formalities. In the case of a meeting with your boss, you *should* spend time preparing beforehand because the meeting will reflect upon your work, your competency, and your role and status within the company. The same could be said for a meeting with your professor to discuss your work as a student.

*Stephen W. Hawking, *A Brief History of Time*. Toronto: Bantam Books, 1988.

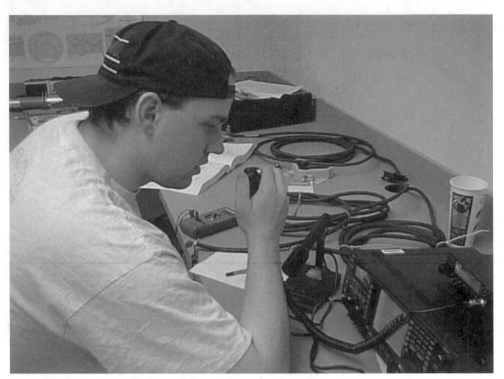

Figure 7.1. Good communication skills are important to all aspects of engineering.

Preparing for an informal meeting takes only a small amount of forethought and planning, because no long speech or formal presentation is required. On the contrary, you may come across as being phony if your conversation seems contrived or orchestrated. Natural speech and hand gestures are always preferable at an informal meeting. You should, however, take the time to think about the content of the meeting beforehand. What will be the topics of conversation? What are your own opinions or thoughts about those topics? Are you being called upon to provide information? If so, take the time to prepare a list that highlights your important points or gives a summary of recent data. Perhaps a one-page outline of your progress to date would be beneficial. Or maybe a list of future planned tasks would be appropriate. Whatever the setting, a brief, to-the-point, one-page document is always helpful at an informal meeting that isn't a social gathering. The following list gives a few examples of informal meetings of various types along with suggestions for a document that might be passed out to all those present at the meeting.

- Project status review: Prepare a one-page bullet list of accomplishments that you've achieved since the last meeting.
- Report on recent tests: Prepare a one- or two-page table showing test results.
- Discussion on market potential of customers: Compile a list of the ten most important customers from the past five years.
- Product design review: Write a one-page summary of the key features of your design concept for the product under development.
- Changes in company procedure: Draw an outline of your proposed organizational chart.

7.3 PREPARING FOR A FORMAL PRESENTATION

Engineers are called upon frequently to make formal presentations to design teams, management personnel, customers, and other large groups of people. Sometimes, engineers must present formal papers at technical conferences. In contrast with an informal presentation, a formal presentation requires careful preparation. You should organize your talk much like a written document. It should contain an introduction, body, summary, and recommendations or conclusion. The following points will help you to prepare an interesting talk that meets its objectives:

1. Know your audience and plan upon an appropriate level of detail.
2. Assume that the customer is hearing about your topic for the first time.
3. Check out audio-visual equipment before the audience arrives. If you are using a laptop computer to present a slide show, determine *ahead of time* that it interfaces properly with the room's computer projection equipment.
4. Dress in suitable attire.
5. Cite the purpose of the talk within the first few minutes.
6. Tell your audience why it's you who is speaking.
7. Show an outline of your talk at the beginning. Give an initial overview of what the presentation will address. A single anecdote can help put the audience at ease and establish rapport.
8. Keep your talk simple. It's easy to get lost in technical details without addressing the main points of your presentation. Leave discussion of details for audience questions. In that way, you'll be able to spotlight only those technical issues of true interest to the audience.
9. Keep your talk short. If you plan to use between 50 percent and 60 percent of the allotted time, you probably will end on time.
10. Ask your own questions, and then answer them. Prepare visual aids to use as your own cues. Use bullets (•) as thought initiators and breakpoints. Use your visual aides, calling attention to them as your talk progresses. Try to format all slides the same way. Maintain eye contact. Never read! If you've brought along notes, refer to them as infrequently as possible.
11. Do not show the audience equations. The audience has little time to decipher them. This rule can be broken occasionally.
12. End your talk with "thank you, any questions?" or a concluding slide. In this way, the audience will know when your talk is over.
13. Restate postpresentation questions so that the entire audience can hear them. Restating a question also will help to clarify its content and will give you an extra moment to formulate your response.

The following example illustrates the elements of a good oral presentation. The context is a student who is presenting the results of recent mechanical loading tests to a group of engineers, students, and professors. The student has been investigating the use of composite materials as part of a high-tech entry for the Peak-Performance Design Competition.

EXAMPLE 7.1: MECHANICAL LOADING AND TESTING

Dan began with a short introduction that explained the purpose of his talk. "Thank you all for coming to my presentation. In this talk I will summarize test results on samples of materials that are candidates for the chassis frame of our Peak-Performance vehicle. As you all know, we've decided to go with composites—matrices of glass, carbon fiber, and epoxy resin—for the principal structural members of the vehicle. This choice will result in some extra costs, because these materials are more expensive than aluminum, plastic, or steel, but ultimately we'll have a better performing and more competitive vehicle. I've done some initial tests on sample compositions and wish to share them with you." Dan displayed the first of his overhead transparencies:

Loading Tests on L-type Carbon Composites
Daniel Little
Peak-Performance Design Competition
Mechanical Design Lab

His next slide summarized the content of his presentation, providing an overview to the audience:

Summary of Presentation

- Description of Composite Materials
- Selection of Test Samples
- Stress-Strain Properties (Nondestructive Test)
- Maximum Load to Failure (Destructive Test)
- Fatigue Tests (Destructive Test)

"Let me begin by providing a brief review of composites. These materials were invented by the aircraft industry as possible light-weight alternatives to more expensive metals, such as titanium and magnesium. Now they're used in everything from bicycles to sailboat masts. Composites are made from a matrix of carbon or glass threads woven into the desired shape and impregnated with high tensile-strength epoxy resin." Dan passed out several samples of composite materials. They were black in color and felt like plastic. He put up a slide that asked two questions he planned to answer during his talk:

Structural Frame for the Peak-Performance Design Competition

- Should we use composites?
- What composition of fiber and epoxy is best?

"Composition, to remind you, is a measure of the percent weight of carbon fiber to epoxy. The diameter of the fibers is also a factor. Basically, the more fiber there is in the mixture, the more expensive the material will be. The trick is to find the optimal composition while considering strength *and* cost. I have here some data on carbon composition, because carbon fiber is the type we're most likely to choose." Dan displayed an overhead that described the various composites he had tested:

TABLE 7-1 Description of Various Composite Materials

PRODUCT	PERCENT CARBON	APPROXIMATE COST PER POUND
L-8	8	$ 96
L-10	10	$ 102
L-12	12	$ 113
L-16	16	$ 120
L-20	20	$ 136
L-24	24	$ 141

"Above about a 24-percent fill rate, this particular composite has too little epoxy resin to hold together well," explained Dan, "and below 8 percent, too little carbon fiber to retain tensile strength."

Dan next described the first of the tests performed on the samples. "The first test consisted of determining the stress–strain relationship for each of the materials in the sample list for a number of different fiber diameters. As a review, let me remind you that *stress* is a fancy word for applied force, and *strain* is another term for the amount by which the material stretches (or compresses) in response to the applied force. In a *linear* material, the strain, or stretch, is directly proportional to the applied stress. Double the stress, and you've doubled the strain. If too much strain occurs due to too much applied stress, the material will go into its nonlinear region in which added stress produces almost no additional strain.

"Now let me explain the tests I've performed. The first involves measuring the stress–strain, or force–displacement, properties of the material using the following setup." Dan put up the overhead shown in Figure 7.2.

"The sample to be tested is first machined into a round bar of 4-mm cross section. It's then installed in a tensile test machine that can apply a stretching force to the bar. The applied force is measured by a load cell that produces a voltage in proportion to the applied force. A strain gage is used to measure the strain. It's actually a thin-film resistor that's glued right onto the side of the test sample. Its resistance changes in proportion to how much the sample has been stretched from its rest length. Here's a typical plot of data taken on one sample, in this case L-8." Daniel put up the slide shown in Figure 7.3.

"The slope of the linear portion of this curve will be equal to the reciprocal of the material's elastic constant. The breakpoint in the curve defines the elastic limit. For our application, we'd like an elastic constant of about 6 kN/mm and an elastic limit of at

Figure 7.2. Instron™ test bed and computerized data recorder. The tensile test specimen is inside.

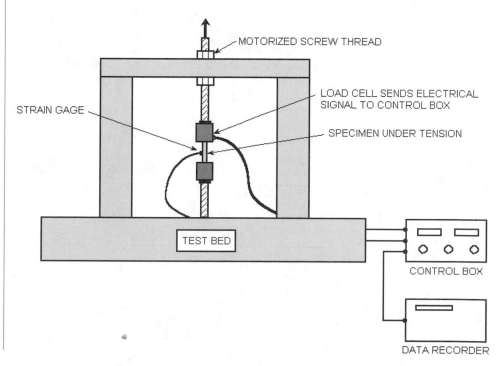

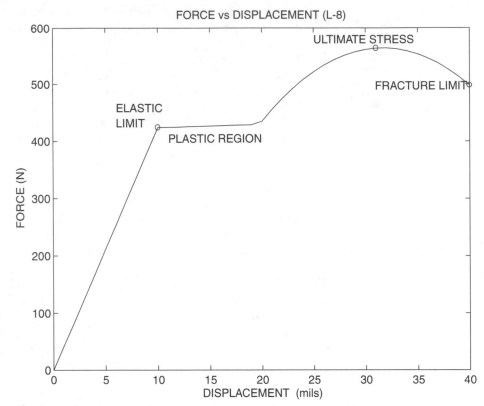

Figure 7.3. Force-displacement curve showing linear region and elastic limit.

least 400 N to ensure an adequate safety margin of about five times the largest force we realistically expect on the components.

"Let me show you the results of measurements obtained from one of my test matrices*." Dan put Table 7.2 up on the screen. He'd copied it from a page in his engineer's logbook, which served as a written record of the data.

"The entries in this first test matrix table indicate the measured elastic constant of each sample in kilonewtons per millimeter of displacement. This next table gives the elastic limit of each sample in kilonewtons." Dan then presented a slide showing Table 7.3.

TABLE 7-2 Measured Elastic Constant (kN/mm) under Tensile Force for 4-mm Diameter Carbon Composite Test Samples

	CARBON FIBER DIAMETER (MILS)				
SAMPLE	3	4	5	6	PERCENT CARBON
L-8	8.4	7.1	6.5	6.0	8
L-10	9.0	7.6	6.6	6.2	10
L-12	9.5	8.2	6.9	6.5	12
L-16	9.7	8.8	7.2	6.8	16
L-20	9.9	9.2	7.5	7.1	20
L-24	10.1	9.7	7.8	7.4	24

*A *test matrix* consists of a table of methodically performed tests in which one variable changes on the vertical axis and one along the horizontal axis.

TABLE 7-3   Measured Elastic Limit (kN) Under Tensile Force for 4-mm Diameter Carbon Composite Test Samples

| | CARBON FIBER DIAMETER (MILS) | | | | |
SAMPLE	3	4	5	6	PERCENT CARBON
L-8	0.28	0.33	0.41	0.52	8
L-10	0.29	0.34	0.42	0.53	10
L-12	0.30	0.36	0.43	0.56	12
L-16	0.32	0.38	0.46	0.58	16
L-20	0.30	0.36	0.44	0.56	20
L-24	0.29	0.34	0.42	0.53	24

In the following figure, I've plotted the most important parameter, the elastic constant, versus percent carbon using fiber diameter as the third parametric variable." Daniel put up the plot shown in Figure 7.4.

"In my next overhead, I've plotted the elastic limit versus percent carbon, again using fiber diameter as the parametric variable." Dan put up the second plot of Figure 7.5.

"Because we need an elastic limit of at least 0.4 kN, we're limited to fiber diameters of 5 mils or more. Similarly, the smallest elastic constants occur for 6-mil fiber, but the 5-mil fiber has adequate properties. Based on my tests, I'm recommending that we go with 16 percent carbon composite with 5-mil fiber for our Peak-Performance vehicle. Thank you for your attention. Are there any questions?"

Figure 7.4.   Force-displacement ratio versus percent carbon.

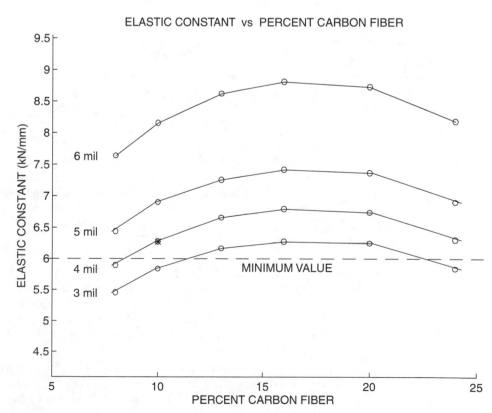

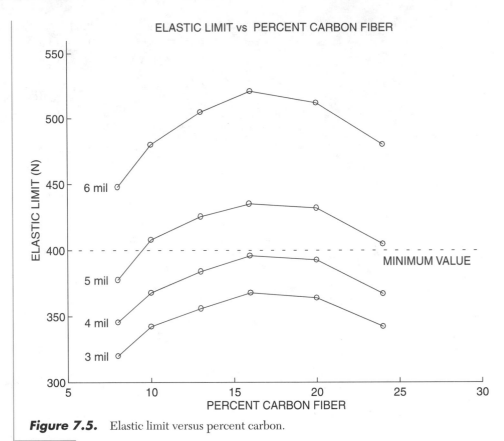

Figure 7.5. Elastic limit versus percent carbon.

7.4 WRITING ELECTRONIC MAIL, LETTERS, AND MEMORANDA

As an engineer, you'll often need to compose short memoranda, faxes, e-mail messages, and notes to fellow engineers, supervisors, employees, or customers. Whether sent on paper or electronically, a memo must convey its message clearly and concisely. It must leave no room for ambiguity or misinterpretation, and it must be written in proper grammatical style. When you engage in a face-to-face conversation, you can modify your communication approach on the spot, depending on the person's reaction. But you will not have the benefit of witnessing the reaction of a person who has received a memo. A written memo must be carefully crafted to elicit the desired reaction on the first reading.

Writing Electronic Mail Messages

Electronic mail, or e-mail, has become the communication tool of choice for countless individuals worldwide. Engineers certainly are included in this group of e-mail participants. E-mail has become an indispensable part of an engineer's work environment as well as an integral part of the design process. E-mail has many uses, but its most common use is as a replacement for hard-copy memoranda. Before electronic mail, communication between employees was conveyed by distributing paper to each recipient. The method was slow and labor intensive.

Electronic mail has made distribution an instantaneous and incidental event. The act of composing an e-mail message, rightfully so, now takes up the largest fraction a sender's time. In the age of electronic mail, the sender has more time to *think* about the what is being written, because almost no time is needed to copy and distribute the document. One of the most common mistakes in e-mail writing, however, is to assume that messages sent electronically do not have to be prepared with the same care as other written documents. An e-mail message that is sloppy and poorly written will not be taken as seriously as one that is carefully written in a professional manner. The wise writer sets a proper tone by preparing e-mail messages with the same care as paper memos. The recipient of an e-mail message will read it in private, just like a written memo, and may treat it with the same formality as a memo received on paper. In addition, an e-mail memo can be copied to large numbers of additional recipients instantly, spawning multiple exposures. Some people print out received e-mail messages, thereby further blurring the distinction between paper and electronic memos.

The formula for writing a competent and effective electronic mail message is easy to learn. A good message should include the following three items:

- A *header* that indicates the recipient, sender, subject, and date,
- A *first sentence* that states the purpose of the message,
- A *body* that delivers the key points of the message.

Header

In our information-abundant world, document organization has become a critical component in business, commerce, and all engineering disciplines. A header that announces the *recipient, sender, subject,* and *date* of the e-mail helps both the sender and receiver to categorize the message and properly file it for future reference. Identifying the subject of the message in the header will also set its tone and prepare the reader for the message to follow. Has the memo been sent to a general distribution list? Is it intended for one person's eyes only? Will the message to follow be formal, informal, alarming, or humorous? A properly designed header will set the tone for the message and prepare the reader to receive it. The header of a memo should look something like this:

To: Karin Peterson
From: Frederick Unlu
Subject: Test Data for Peak-Performance Motor Evaluation
Date: April 12, 1999

An alternative form suitable for a message sent to a group of people might look like this:

To: Distribution
From: Tina Oulette, Team Leader
RE: Next Team Meeting
April 12, 1999

Note that many electronic mail systems include automatic prompts for header entries, and most all include a date stamp on the message regardless of whether the sender inserts one on his or her own. Nevertheless, it's a good idea to specifically include a header similar to one of the foregoing examples, because the Internet system often adds network routing information between the automated header and the body of the text.

First Sentence

A good message will clearly state its purpose in the first, or possibly the second, sentence. A sentence that begins with, "I am writing this message to inform you . . ." will help the reader unambiguously understand why you have written the message. Stating the purpose of the message right at the beginning will help formulate its tone and state its objectives. Will your message provide information, request a response, ask for permission, or give instructions? The structure of the first sentence will determine the way in which the body is received. It will also ensure that the reader does not misinterpret your reason for writing.

EXAMPLE 7.2:
THE TRIP
REQUEST

As an example of the power of the first sentence, consider the following memo which was sent by an employee of an engineering company. The engineer was writing to his boss to ask for permission to attend a conference of the American Society of Mechanical Engineers.

> To: Roscoe Varquin
> From: Harry Coates
> Subject: Upcoming ASME Conference
> Date: January 14, 1998
>
> Roscoe,
>
> As you may know, the American Society of Mechanical Engineers is holding a conference on lightweight composites in Dayton, Ohio, at the end of June. I think that someone from our company should attend this meeting. Over the past several years, composites have shown promise as a viable alternative to steel or aluminum. They combine the strength of the former with the light weight of the latter. It's time that we learned more about these important materials.
>
> The conference will be held at the Dayton Buena Vista. I've spoken with our travel agent and found that flights will cost about $300 for a round trip. Hotel will be $82 per day, and the conference registration fee is $180. Please let me know what you think.

Critique   Harry began his message by explaining the upcoming conference and proceeded to present his data about how much it would cost him to make the trip. Despite what seemed like a clear explanation, however, Harry never got to go to the conference, even though Roscoe was convinced of its value to the company. Harry presented his data and did his research but never stated explicitly in his first sentence that it was *he* who wanted to attend the conference. The message brought the matter to Roscoe's attention, and it was Roscoe himself who wound up going to the conference! The confusion resulted because Harry failed to clearly state the purpose of his memo. His e-mail message would have been *much* more effective had it begun with the simple sentence, "I would like permission to attend the upcoming ASME conference in June." Harry learned, much after the fact, that it's important to state the purpose of a memo in its first sentence. Had he talked to Roscoe in person, he might have detected the confusion by witnessing his boss's reaction on the spot, but his chosen method of communication omitted the link of personal contact.

Body

When composing the body of an e-mail message, follow basic rules of style and grammar. Each idea or concept should have its own paragraph, and paragraphs should never consist of a single sentence only. Each paragraph should follow logically from its predecessor so that the flow of ideas makes sense to the reader. When you compose a message, think about the sequence of ideas so that your message has structure and flows logically. Use the full-screen edit feature of your software, rather than the one-line-at-a-time mode available on some e-mail systems. In this way, you will be able to go back and revise or restructure the body of the message before sending it.

EXAMPLE 7.3: VISIT TO THE CUSTOMER'S WORKPLACE

One quality of a good e-mail message is its ability to convey correctly all the subtleties of a given subject. Without personal contact to enhance communication, an e-mail message must be precise and readily comprehended on the first reading. Consider the following not-so-well written message that describes the visit of a software engineer to the client's workplace. Try to identify the deficiencies in the memo. Its subject concerns a program called the Universal Information System. The task involved rewriting an original UNIX version of the program to run under Microsoft Windows98. The software program is used by the client to keep track of customer charge card records. The software engineer in charge was writing a memo to her boss summarizing the results of a recent exploratory visit to the client.

> To: Roscoe Varquin
> From: L. Berkin
> Subject: Universal Information System
> February 26, 1999
>
> On the 25th of February, my group met with the customer. We were allowed to use a user's ID in order to bring up his account record on the UIS. The customer crossed out the user's name on the sheet containing his information, but it turned out that the UIS listed his account with both his name and social security number. My group made a record of how the system was set up in order for us to have the same heading in our program.
>
> I learned some of the commands that are used on the UIS Galaxy system. The command TR14 finds the user's in-question account and displays it on the monitor. The TR33 command displays all transactions which the user made since the specified date. Last, but not least, the TR35 command displays all payments made since the given date. It turned out that this command was a recent one designed and implemented by one of our fellow employees who recently left the company. In our program, we will assume that this command will be used by the customer, even though there is doubt that no one would keep it up to date except for the fellow employee who's stay there is not guaranteed.
>
> Meeting with the customer at her office made it clearer what they expect of the product. They want the product easy to use. They want the account summary sheet to highlight all the transactions not yet paid for. In all, I learned several commands and got some exposure to the UIS system. This was a profitable meeting.

Critique   Despite its poor grammar and flow of ideas, the body of the message does contain the following key points:

- Opportunity to see demonstration of UIS system
- Customer erroneously allowed us to see the identity of a sample account holder.
- Program within UIS system is called Galaxy

- UIS system includes several commands:

  -TR14: Customer account

  -TR33: Transactions display

  -TR35: Payment history

- TR35 command designed by previous employee of company
- Meeting was useful.
- Customer wants unpaid transactions highlighted.

Regardless of its content, however, the body is deficient, because the order in which the information is presented is entirely random. The message lacks logical progression, contains no preamble, and includes no added comments to aide in interpretation. Each fact is presented with no accompanying explanation of how it fits into the overall context of the memo. Additionally, the information contained in the first paragraph is irrelevant to the writer's boss who would not care that the customer made a mistake while attempting to access the system. The writer seems to have had a list of facts in her head and proceeded to regurgitate them on paper in whatever order they came to mind. The message also omits critical information. The author of the memo does not mention the identity of the customer or who the person was that gave the tour and demonstration. There is no mention of what the letters "UIS" stand for, and the term "Galaxy" is used without explanation. It's highly likely that this memo underwent very little onscreen editing or rewriting and probably was poorly received.

Now consider the following version of the message which was written and revised using proper grammar, construction, and form prior to being sent.

To: Roscoe Varquin

From: L. Berkin

Subject: Summary of customer meeting at Boulton Industries

February 26, 1999

Roscoe,

This memo summarizes what we learned during our visit to Boulton Industries on February 25, 1999. Our project team met with Boulton's representative, Ms. Connie Donaldson, at their Weston office. The meeting provided us with an overview of the customer's existing Universal Information System (UIS) and introduced us to the procedures commonly used in Ms. Donaldson's office. We worked with her on the UIS system using actual account records.

The UIS system contains an application program called Galaxy that allows the customer to view account records. The record of an account holder is accessed by entering an ID number into the Galaxy system at the menu prompt. The account data can then be viewed in a variety of formats useful to the customer.

Several user programs, called Transaction (TR) screens, reside within Galaxy and help the user to display account information. These TR screens are used routinely by Boulton employees for processing monthly bills and servicing customer call-in inquiries. During our meeting, we recorded the display headings of each important TR screen so that we can create the same headings on our PC-based Windows98 version of the UIS system.

As part of our tour, Ms. Donaldson demonstrated some of the commonly used TR screens. One command, for example, is called TR14. It retrieves a specified account record and displays it as a standard mailing form on the monitor. A second command, called TR33, displays all transactions that have occurred since a date specified by the user. A third command, called TR35, displays the history of all payments made since the user-specified date. This last command was written previously for Boulton by one of our employees. When writing our new version of the program, we should assume that this command again will be used by Boulton to process records and should plan to integrate the feature into the Windows version of the program.

I feel that we learned enough from our meeting to enable us to proceed with the Boulton UIS project. If necessary, we can return to Ms. Donaldson's office at a later date to obtain more information.

Sincerely yours,

Laura Berkin
Software Division

Writing Formal Memos and Letters

The informality of an e-mail message in not appropriate for all communication. At times, the added formality of a written, hard-copy letter is preferable. Formality suggests importance. Applying for a job or sending a follow-up thank-you letter, for example, are situations that call for a formal written letter. Likewise, if your information has archival quality, or if your message may have legal implications, then you should send your memo in paper format. Besides carrying more social weight than an e-mail message, a paper letter can bear a binding signature.

The rules for composing and sending a written letter are almost identical to those for sending electronic mail. One key difference is that a formal letter normally does not contain a To and From header, but instead begins with the recipient's address letter and a formal salutation. A letter should be well presented on good paper or letterhead and be printed in an attractive format. The following example illustrates some of the finer points of writing an effective formal letter. The first version shows the letter as originally written, and the second shows the results of a much needed revision.

EXAMPLE 7.4: SUMMARY OF TESTS RESULTS

The following letter was written by an engineer who wished to summarize the results of mechanical loading tests for a client. The letter is based on the following points, listed here in the order in which the author wrote them down in his logbook:

- Initial loading tests are completed
- Test samples were composites of carbon and epoxy
- Control samples were steel
- Same shape chosen for each; steel was machined, composites were molded
- Samples were composites and steel
- Initial difficulties fitting samples into test machine
- Made a holding jig to allow testing
- We should go with composites at slightly increased diameter
- Numerical data from test results:

Diameter	Composite	Steel
0.25 in	245 lbf	321 lbf
0.375 in	1644 lbf	1790 lbf
0.50 in	3021 lbf	3229 lbf

The letter as written by the engineer reads as follows:

Apex Systems
Structural Testing Laboratory
730 Commonwealth Ave., West Roxbury, MA 02132

Helen Brickland
Access Engineering
44 Cummington St.
Boston, MA 02215
January 18, 1999

Dear Ms. Brickland:

We've completed the initial loading tests on the samples made from composites and steel. The samples had the same shape, and the steel was machined and the composites molded. We had some initial difficulties fitting the samples into the test machine but finally made a holding jig to allow testing. I think that we should go with composites at slightly increased diameter.

Here are the pieces of data:
  0.25 in. diameter: composite 245 lbf, steel 321 lbf
  0.375 in. diameter: composite 1644 lbf, steel 1790 lbf
  0.50 in. diameter: composite 3021 lbf, steel 3229 lbf

Sincerely yours,

Ed Garber
Mechanical Engineer

Critique   This letter is correctly formed and includes all needed information, but the writer would have done a better job of stating his objectives if the first sentence had cited the real purpose of the letter, which was to make a recommendation to Ms. Brickland that composites be used instead of steel. A better opening might have been the following:

Dear Ms. Brickland:

For the past month, Apex Systems has been performing loadings tests on samples of composites and steel for Access Engineering. Based on our test results, I'd like to recommend that we choose composites for this project.

The original letter has other problems beyond those of its first sentence. Its body, for example, is completely disorganized. Although the author has followed almost verbatim the ordering of ideas as recorded in his logbook, those ideas were not particularly well ordered to begin with. The letter reads disjointly, as if it's been poorly edited. The sentences are choppy and do not flow from one to another. The numerical data should have been presented in more concise, tabular form. Finally, the author failed to describe the purpose of the tests, didn't go into detail about how they were performed, and didn't describe the test samples other than to say that they were a mix of composites and steel. He should have provided the dimensions of the samples, the composition mix of the composites, and how many of each type of sample were tested.

A revised version of the letter that corrects for these deficiencies might look something like the following:

Apex Systems
Structural Testing Laboratory
730 Commonwealth Ave., West Roxbury, MA 02132

Helen Brickland
Access Engineering
44 Cummington St.
Boston, MA 02215
January 18, 1999

Dear Ms. Brickland:

The mechanical group working on the Delta vehicle project with Access Engineering has just completed tests on samples of the composite and steel materials that we are evaluating for the main structural members of the frame. Based on our test results, our group recommends that we choose composites of slightly enlarged diameter for the structural materials. The details of the tests are described below.

Samples of machined steel and molded composites were fabricated in the shape of standard tensile test specimen bars having a variety of diameters in the range 0.25 in. to 0.625 in. (ASME specification 246). These samples were stressed to the breaking point in our lab's Instron test machine. Although we had some initial difficulties in fitting the samples of diameter other than 0.5 in. into the test machine, we were able to make a holding jig to accommodate testing of all samples. The numerical data from our tests results are summarized in the following table:

BREAKING FORCE UNDER TENSION				
Diameter (in):	0.250	0.375	0.500	0.625
Composite (lbf):	845	1644	3021	4229
Steel:	1421	2790	4310	5541

As you can see, our minimum targeted breaking strength of 4000 lbf can be met with either a composite rod of 5/8-in (0.625") diameter or a steel rod of 1/2-in (0.500") diameter. Given the much lighter weight of the composite material, however, its strength-to-weight ratio is much higher than that of steel. I recommend that your company go with composites for this project.

Sincerely yours,

Ed Garber
Mechanical Engineer

This second version is much improved over the original. The author has articulated the purpose of the letter in the second sentence, rather than in the first, but it works well here because the letter still reads and flows nicely. The first sentence serves as a preamble to the real purpose of the letter which is revealed in the sentence that begins, "Based on our test results . . ."

The data are also well presented in the table contained within the letter. Instead of putting the units next to each entry, Ed has written them just once next to the category titles on the left-hand side. This format makes for a much neater display of data. Ed also has elaborated on the details of the tests. When Ms. Brickland reads the letter, the context in which the tests were taken will immediately be clear. This point is an important one. Ed might falsely conclude that Ms. Brickland will understand its context, because his primary work assignment over the past month has been the taking of data and the testing of the materials. But Helen, as a manager, is likely to be juggling dozens of projects and details, and she will welcome a letter that first refreshes her memory about its context and background. In addition, she may find herself reading the letter at some future time when the background details of the test procedures have faded from memory. Or she might forward the letter to someone else who is unfamiliar with the details that Ed has correctly provided.

PROFESSIONAL SUCCESS: PREPARING A PRESENTATION FOR A NON-TECHNICAL AUDIENCE

At times an engineer must make a presentation to a non-technical audience. You also may encounter this challenge while you are a student. Perhaps you'll be invited to speak at your old high school. Maybe you'll be asked to explain the activities of a research laboratory to visiting parents and prospective students. Or maybe you'll receive a class assignment to prepare a talk on a technical subject for a general audience. In these and similar situations, the following few basic principles can guide you:

- Assume the audience knows nothing about your topic.

- Explain background material without using jargon words. (You'd be surprised at how many seemingly common words are really part of the engineer's private lexicon.)
- Start at the beginning to provide the big picture.
- Pretend that you're speaking to a group of fourth graders.
- Never show equations to a non-technical audience. (You'll probably pique their math anxiety.)

7.5 WRITING TECHNICAL REPORTS, PROPOSALS, AND JOURNAL ARTICLES

Technical reports, proposals, and journal articles are the engineer's equivalent of term papers. As an engineer, you are likely to face the task of writing one of these documents at some point in your career. Unlike short e-mail messages, memos, and letters, which are usually only a few paragraphs long, longer technical documents require considerable thought and preparation and usually cannot be finished in one sitting. The sections that follow highlight some of the key elements of these longer documents.

Technical Report

A technical report is used to convey important findings or test results to a controlled audience. Technical reports seldom undergo peer review, and distribution of the report is done at the discretion of the author or employer.

The typical technical report is between two and twenty pages long. In content and form, it's not unlike a lab report you might prepare in some of your college courses. Although the format may vary somewhat, most technical reports contain the following elements: Introduction (or Background), Experimental Setup (if applicable), Theory, Data, Analysis, and Conclusion.

The introduction serves as a preamble to the document and states its reason for having been written. Unlike the first sentence of a simple memo, the introduction of a technical report can occupy a paragraph, a page, or sometimes many pages. A person pressed for time will merely skim the introduction, so it, along with the conclusion, should provide a self- contained overview of the entire report.

If the document describes the results of an experiment, an experiment section should be included that describes the physical setup. This section should provide enough detail that a reasonably competent person could completely reconstruct the experimental apparatus and obtain similar results. It should describe instruments, apparatus, mechanical techniques, dimensions, and other key parameters.

The data section includes the results of any experiments or tests that were performed. It should explain why each set of data is presented, how it was obtained, and what bearing it has on the main purpose of the document. A report or journal paper is likely to be used later as a reference source, so it's important to present data completely and accurately. The presentation should be easily digested by someone not intimately familiar with the details of the project.

The analysis section is where the data are evaluated, interpreted, and used to support any claims made in the report. Mathematical calculations belong in this section, as do plots and charts derived from the data. In some cases, particularly in reports that deal with design work, the analysis and data sections appear in reverse order. First the analysis of the device is presented, followed by data on tests that show whether the device meets the expectations or predictions of the analysis.

Finally, the conclusion is used to summarize the claims, results, and observations included in the report. Some individuals may not have time to read the whole report, but need to be familiar with its content. The conclusion section should be written to serve the needs of a person who only has time to browse or skim the report. The conclusion should be a stand-alone section that summarizes all the key points of the report.

Journal Paper

Journal papers provide a way in which engineers disseminate information of interest to other engineers. They also provide a public forum in which engineers and scientists can announce "first to invent" status of a new technology or discovery. Most journal papers, particularly those sponsored by professional organizations, must undergo peer review before publication. This procedure helps insure their quality and accuracy. Although the standard format—introduction, theory, experiment, data, analysis, and conclusion—is appropriate for journal papers, many publications specify their own format in which a journal paper must be submitted.

Proposal

A proposal differs from a report or journal paper in that is its primary objective is usually to secure *money*. A proposal attempts to convince a client or funding source that your organization can best handle a research or design job, or perhaps that your product will be best for a particular application. In addition to the various sections of a technical report, a proposal often includes additional sections on objective, budget, company background, and personnel.

7.6 PREPARING AN INSTRUCTION MANUAL

One of the most common documents written by engineers is the instruction manual. An instruction manual introduces the user to your product and provides information regarding its setup, operation, and use. A well-written instruction manual also includes sections on safety information, troubleshooting, repair, and theory of operation, if applicable. While not all engineered devices require an instruction manual (the operation of a snow shovel, for example, should be self explanatory), those that involve detailed operating procedures should be accompanied by instructions. Indeed, user perception of many products is derived directly from the quality of the instruction manual.

The sections of a typical instruction manual are outlined below. If the manual is long, a table of contents with page numbers should be included. Obviously, the need for each of the sections suggested below will depend on the specific product described by the manual.

Introduction

The introduction should provide an overview of the product. It should explain the purpose of the product, its usefulness to the user, its special features, and proper use of the manual itself. Various titles for the introductory section include "Getting Started," "Welcome to Product X," "Before Using Your Device," etc.

Setup

The setup section should outline the procedures that the user must follow before the product is ready for use. This section should appear at the front of the manual, where it will be easy to find. It should guide the user through the setup procedure step by step, and it should use illustrations liberally.

Operation

The operation section of the instruction manual is its most important part. A first-time user will use this section to learn how to operate the product and will refer to it thereafter to clarify points of operation. Because of its likely use as a reference source, the operation section should be carefully laid out and organized so that the user can extract information without having to read the entire manual from the beginning.

Safety

Safety is a very important aspect of engineering design, and any such information relevant to the well-being of the user must be included in the instruction manual. If the product has dangerous moving parts or high voltages, includes safety panels or guards that must not be removed, or has the potential to emit flying objects or capture loose clothing, then appropriate warnings should be included. In our overly litigious society, where personal injury lawyers advertise freely on the radio and TV, safety warnings have become pervasive. Some safety warnings may seem overly cautious or even ridiculous (e.g., don't put your fingers in the moving blades or you might get hurt), but safety warnings have become a necessary part of engineering.

Troubleshooting

Expect your product to break down, regardless of how well it is made. If it should happen not to malfunction over its lifetime, consider yourself lucky. Designs do fail, and the troubleshooting section should guide the user through a simple set of tests that can help identify the source of any malfunction and get the system running again. The section should outline simple repairs that the user can try before taking the more drastic step of returning the product. If appropriate, include a section that explains how to get in touch with the manufacturer (you) in case difficulties cannot be resolved by the user.

Never assume that the user will understand something that is not clearly specified. The following troubleshooting entries may seem silly, but many instruction manuals contain similar versions:

Symptom: No lights or displays of any kind are lit; unit appears to be dead.
Possible Cause: Unit is unplugged or has blown fuse.
Remedy: Plug cord into proper outlet. Replace fuse.

Symptom: No sound coming out of speaker.
Possible Cause: Volume control is turned all the way down.
Remedy: Turn volume knob clockwise.

Symptom: Drive shaft does not turn.
Possible Cause: Clutch is not engaged.
Remedy: Engage clutch by moving lever to on position.

Symptom: No flame from burner.
Possible Cause: Pilot light extinguished.
Remedy: Relight pilot. (See Section 2.1: Lighting the Pilot Light.)

Symptom: Pilot light cannot be lit.
Possible Cause: No gas supply.
Remedy: Open main gas valve; replace propane tank.

Appendices

Information likely to be of interest only to a special readership should be included in appendices. Examples include circuit schematics, exploded assembly diagrams, theory of operation, and lists of part numbers.

Repetition

One of the subtle features of a good instruction manual is its ability to engage the reader regardless of where he or she begins reading it. This attribute can be achieved by repeating informative details at several junction points throughout the document. As you write the manual, imagine the viewpoint of a reader who has begun to read it somewhere in the middle of the document. Restate key information at topical transitions and major section headings, rather than referring to previous sections. Do not assume that the reader remembers details covered previously if they are relevant to the present section.

**EXAMPLE 7.5:
THE ATM
SIMULATOR**

The following example contains (slightly edited) excerpts from an instruction manual written by senior project students in the Department of Electrical and Computer Engineering at Boston University°. It includes many of the elements mentioned above and illustrates the overall features of a good instruction manual. The manual explains the operation of a simulator for a bank automatic teller machine designed to teach banking procedures to elementary school and special needs students.

° "G. DeBernardi, R. DeMayo, M. Givens, M. Magne, E. McMorrow, and S. Tansi, *Automated Teller Machine Simulator Instruction Manual*. Terriers Technologies of Boston, 1992

Automated Teller Machine Simulator
Instruction Manual
Terrier Technologies of Boston

WELCOME TO THE TTB ATM SIMULATOR
The Terrier Technologies of Boston Automated Teller Machine simulator has been designed to help you teach students the important aspects of banking skills. The unit has been designed for simplicity and ease of use. It should provide you with many years of trouble-free operation.

HOW TO USE THIS MANUAL
The TTB ATM users' manual is divided into five sections and an appendix. The first two sections provide a general overview of setup and system start up, respectively. The third section explores the inside of the ATM simulator and discusses its various modules. The remaining two sections deal with care and troubleshooting of the equipment. The appendix contains wiring diagrams and computer software codes.

OVERVIEW OF OPERATION
The TTB ATM simulator is a self-contained product. Each of its modules simulates the operation of a real bank ATM machine. Account information is stored in the computer connected to the ATM panel. A banking session begins by prompting the user to insert a card into the card slot. Once the card has been properly inserted, the user is asked for a password to be entered via the keypad. After entering the correct password, the user is asked to choose from the following list of transactions:

1.   Deposit
2.   Withdrawal
3.   Fast Cash
4.   Account Balances

Once the type of transaction has been chosen, the simulator asks the user to choose between a savings and a checking account. The user can withdraw facsimile money from the cash dispenser or place an envelope in the deposit slot. Immediately following the transaction, a receipt of the session is printed out and the user's account is updated inside the computer.

The system operator has additional choices not shown on the main user menu. These additional choices include: `Modify Parameters`, `Troubleshooting`, and `Print Account Information`. The `Modify Parameters` command enables the system operator to update user accounts. The `Troubleshooting` command tests the individual modules of the ATM simulator. `Print Account Information` makes a printout of all user names and account balances. The ATM simulator is also equipped with two flip-up panels. The top panel exposes the main keyboard used by the system operator to initiate and update accounts. The bottom panel provides access for paper replacement for the receipt printer and provides access to the cash dispenser.

INITIAL START-UP (SYSTEM OPERATOR)
The power switch is located on the rear of the unit. Plug in the power cord and move the power switch to the ON position. You should hear a whirring sound as the internal disk drive is activated. To begin the session for entering or updating user accounts, first access the keyboard by raising its hinged cover. The screen display will give instructions on how to proceed.

SETTING UP OR CHANGING AN ACCOUNT

To set up a new account or change a previous account, access `Modify Parameters` from the `Main Menu`. Once this selection has been made, the following set of selections will appear on the screen:

1. ID Number
2. Password Savings
3. Checking
4. Account Balances

Choose your selection by pressing the appropriate number on the numeric keypad. Using the arrow keys, move the cursor to the entry for the user account to be edited. The following set of commands can be used to enter and edit field data:

`Enter:` Access the field that needs to be added or edited.

`Ins:` Add information to a user field

`Del:` Delete information from a user field

`Esc:` Return to the Main Menu

All fields must be filled if the system is to access properly each user account. Once all account information has been entered, a full printout can be made by choosing `Print Account Information` from the `Main Menu`.

BEGINNING A BANKING SESSION

When all user accounts have been entered, the unit is ready for simulated banking sessions. From the `Main Menu`, select `Begin Session` to begin a simulation. The screens will guide the user through the entire process in a manner very similar to a real automated teller machine.

INSIDE THE ATM SIMULATOR

An IBM 8088 computer is used to control all aspects of operation for the simulator. The system operator controls the account balances and all other program functions via the keyboard. The display consists of a 13-inch monochrome monitor. The screen guides the user through the entire process and provides help whenever necessary. Display screens are similar to those found on real ATM bank machines. Additional screens take the system operator through the account information sequences.

KEYPAD

The TTB ATM keypad is identical to that found on real Diebold ATM machines. It consists of fifteen keys (11 blue and 4 white.) The 0 through 9 blue digits and decimal-point keys are used for selecting dollar amounts, and the four white keys, labeled A, B, C, and D, are used for making transaction decisions. Pressing the CANCEL key will terminate the session at any given time.

CARD SLOT

After the system operator has set up user accounts, the simulator will wait for the insertion of a bank card. The card must be placed into the machine in the proper direction shown on the diagram on the ATM front panel. The card will be pulled into the slot by the TTB ATM and will remain in place unless the card is inserted in the wrong direction, the transaction is terminated by the user, or the system is shut down or loses power.

MONEY DISPENSER

The user may ask the ATM for a withdrawal in increments of ten dollars. If the user asks for cash in other increments, the simulator will inform the user that the transaction is not allowed and will suggest trying again. Once an amount has been specified for withdrawal, the cash dispenser will drop facsimile ten-dollar bills into the money bin. The system operator can reload the machine with money when needed.

RELOADING

When bills need to be reloaded, the operator should access the cash dispenser from the back of the TTB ATM by pushing down on the springs and placing a neatly packed stack of facsimile bills into the unit. Be sure that the bills can roll out freely by manually turning the dispensing wheels until the bills can move easily.

DEPOSIT SLOT

As in most ATM machines, money or checks can be deposited by placing a deposit envelope into the deposit slot. After the simulator has informed the user that it is ready to receive a deposit, it awaits the insertion of an envelope. The deposit slot module will pull the envelope inside the unit. Deposit envelopes may be collected later by the system operator from the deposit bin. An indicator light located on the back panel of the TTB ATM will inform the system operator if a deposit envelope has been received. Once the envelope has been securely received by the ATM, the user's account will be updated automatically.

PRINTER

After a transaction has been completed, the simulator will produce a copy of the transaction and balance information. The printer is mounted on the side of the unit for easy access. The paper is the same type used in adding machines. Its roll should be placed on the bar so that it can be fed into the printer. When the paper roll runs out, it can be easily replaced by sliding a new roll on the bar. The printer is connected to the computer via a standard Centronics printer interface cable.

EQUIPMENT MAINTENANCE

The main unit may be cleaned with a damp (not wet), lint-free cloth. Aerosol sprays or other cleaning solvents are not recommended for cleaning. If the screen gets dirty, use a clean cloth or paper towel to wipe the screen.

The unit contains no user serviceable internal parts. Repairs should be performed only by a qualified TTB technician. Moving or tampering with any of the components of the TTB ATM may adversely affect the operation of the overall system.

SAFETY PRECAUTIONS

Even though the TTB ATM unit is designed to be as safe as possible, a few circumstances can lead to hazard conditions. For your own safety, and that of your equipment, always take the following precautions. Disconnect the power plug if . . .

- the power cord or plug becomes damaged.
- any piece of clothing gets caught in the bank card, printer, envelope, or money dispensing slots.
- any liquid is spilled on the unit.
- the ATM simulator is dropped.

TROUBLESHOOTING

Problem: No lights or displays of any kind are lit; unit appears to be dead.
Possible Cause: Unit unplugged or has a blown fuse.

Problem: Keys are stuck.

Possible Cause: Keys may begin to stick due to temperature or excessive use. Try to loosen stuck keys by gently pulling them up. WARNING: Do not pull too hard or the keys may break off.

Problem: No response from keypad.

Possible Cause: The TTB ATM may not be in transaction simulation mode.

Problem: Envelope is not being pulled into the deposit slot.

Possible Cause: The wheel of the drive motor may be stuck. Pull up on the traction wheel or remove the envelope and push it manually through the slot.

Problem: Bills are not being dispensed from the machine.

Possible Cause: The money dispenser may be empty.

SERVICE AND SUPPORT

Service for the ATM simulator is available through our local field service network. Please contact:

Terrier Technologies of Boston

ECE Department, Boston University

8 Saint Mary's St.

Boston, MA 02215

617-353-9052

7.7 STRATEGY FOR PRODUCING GOOD TECHNICAL DOCUMENTS

With the possible exceptions of the short memo and e-mail message, writing good documents takes preparation, time, and effort. Whether your writing task consists of an instruction manual, technical report, journal paper, annual summary, or an entire book, it will stand a better chance of accomplishing its objectives if it is well written. As with any other skill, learning to write well requires practice, patience, and attention to detail. In this section, we review several time-tested techniques for improving your writing abilities. Although different writers develop individual styles, most follow the same basic rules outlined below.

Plan the Writing Task

Before sitting down to write, gather all pertinent information. Assemble the results of design calculations, tests, experiments, user specifications, and all other available material. If it's pertinent to the writing task, make sure your engineering logbook is by your side. Gather reference citations, figures, and graphics, if applicable. Have everything in your disposal before you begin the writing task.

Find a Place to Work

One of the most important lessons to learn about writing long documents is that you must devote an uninterrupted block of time to the job. It's simply not possible to write well if you are distracted by phones, e-mail, television, or people coming to talk to you. Writing a complex document takes a long time, sometimes hours or days. When you face a writing task, persistent concentration over an extended period of time will help get you into a creative writing mode. Your mind must sort ideas, arrange their flow, and commit them to well-written and enticing prose. Choosing precise words is an art form similar to painting, sculpting, or composing music. Your writing will flow more smoothly if you find a secluded spot where you'll have absolutely no interruptions. The telephone, your e-mail terminal, or other people stopping by will almost certainly break your concentration during writing and interrupt the creative process. In an ideal world, writers would be able to post "Writing in Progress: Do Not Disturb" signs over office doorways or cubicle portholes. In the real world, such sequestered spots are not always available. Go wherever you can find privacy. A library, cafeteria (during off hours), lab bench, or even the corner donut shop provide mental seclusion, despite their public nature, because they provide uninterrupted time for writing.

Define the Reader

Decide who will be reading your document. Some readers will know more about your subject than you do. Others will know nothing at all. It's important to know the technical level of your reader so that you can set the *tone* of the document. Suppose, for example, that you are reporting on Peak-Performance loading tests for a group of nonengineering students. In such a case, you probably wouldn't include material on spring constants, test methods, or Young's modulus of elasticity. If the report were for engineering students or professors, you might want to include these items.

Regardless of the technical level of your readers, you should decide how much detail the reader will need about the topic. Also be aware of what the reader will do with the document. Will it be redistributed? Will someone else read it? Answering these questions will help you set the tone of the document.

Make Notes

Professional writers always seem to get their documents to read just right. You, too, can produce well organized, easy-to-read documents by mastering one valuable method used by the professionals. Before you begin writing the actual document, make random, stream-of consciousness notes—one line reminders—of *anything* that might need to go into the document. At this stage, give no particular attention to order or emphasis. Include the obvious essentials, as well as the possibly needless trivia. Many writers find it more effective to perform this step with traditional paper and pencil rather than on a computer, because the act of keyboard typing is known to occupy a sizable fraction of brain activity, leaving less mental power for creative and organizational activities. Regardless of which method you choose for recording your ideas, the key at this stage is not to worry about the order in which you write things down on your list. Commit all your ideas to paper now. You'll scrutinize them at a later step in the writing process.

Create Topic Headings

Your next step should be to form the overall structure of the document. To accomplish this task, you should write down the topic headings that will need to go into the finished work. Again, you should write these items down in random order, paying no attention at this stage to how they will be structured. Each of these topical headings will eventually become a paragraph or sequence of paragraphs in the finished document. When you're done with your list, examine each topic heading to see if additional headings come to mind. Delete irrelevant headings and group remaining headings into the main topic areas of the document.

When your list of topic headings is complete, arrange them in a suitable order. Decide which order of presentation is the most interesting, logical, and easiest to understand. It's at this point that the main structural framework of the document begins to take shape.

Take a Break

Before you begin to write the actual document, take a break. Clear your mind before beginning the writing process.

Write the First Draft

If you've done a good job of preparing your list of one line notes, you'll be ready to begin the writing process. Find an interruption-free spot to work, and start to write. Don't worry about writing in perfect form at this stage. Expect to revise your document many times before it's completed. The important thing during the first draft stage is to get your words down on paper. Use a word processor or write by hand and type later, whichever method suits you. As previously mentioned, using a keyboard detracts from the creative energy of writing for some people, so don't feel that you must compose the first draft on a word processor. A word processor is an indispensable tool for writing, of course, but it's main advantage comes in the revision process. On the other hand, if you *can* learn to compose the first draft directly on a word processor, you'll save a lot of time otherwise spent in transcribing handwritten pages.

At this point in the writing process, don't be too concerned about spelling or exact phrasing. These aspects of the document will be corrected and modified later, in the revision phase.

If the work is short, write the first draft in one work session. If the document is long, divide the work into medium-length work sessions. At the end of each work session, rapidly scan the draft and make only *obvious* changes. Do not do major revisions at this time. When the draft is finished, again take a break to clear your mind. If time permits, set aside the document for another day so that you can approach it with a fresh perspective.

Read the Draft

After your break, when you are no longer intently focused on the document, reread it as if you are seeing it for the first time. Check your writing style for clarity. Are there vague, confusing, or ambiguous passages? Are the sentences in the correct order? Is there a logical flow within each paragraph and between successive paragraphs? Check for correct tone. Is the writing style suitable for both the subject matter and the reader?

Try not to read the draft solely from your computer screen. A document always looks different when it's been printed out on paper, because the reader is able to absorb more text at once and get a better feeling for how the entire document is structured.

Revise the Draft

A document seldom is ready for distribution after the writing of its first draft. After you write the first draft, take the time to review your work. Revise words, reword sentences, rearrange paragraphs, and reorganize sections to further refine and clarify meaning. As you revise, mercilessly slash unnecessary words and sentences. Weigh each word and phrase, and keep only those words and phrases that carry important meaning. Technical writing should be direct and to the point. Keep each paragraph relevant. Replace complicated phrases with simple words, and limit superlatives. As you reread your document, give it at least one pass in which you ignore content and look at words, sentences, and paragraphs in their grammatical context only. Remove "fat," unnecessary words, and details that have low information content. You should also recheck factual statements, formulas, numbers, and calculations for accuracy. Be careful to proofread material cited from other documents. Most all word processors include a spelling checker. Get into the habit of using one before sending out a document of any kind.

Revise, Revise, and Revise Again

After you've made your first revision, revise, revise, and revise again. Most good writers devote three or more, and sometimes dozens, of rewrites to get a document to read just right. Each chapter of the book you are reading now was revised at least six times before being sent to the editor for publication.

Review the Final Draft

When you feel that your document is finished, put it aside and come back to it, preferably on another day when you will not be prejudiced by the intensity of the writing process. As you read your document, evaluate it as an outside reader would. Keep an open mind and ask yourself the question, "How would *I* react to what I have written? Will it produce the intended reaction or response from the reader?" If the answer is "yes," your document is ready for the outside world.

Common Writing Errors

Errors in usage and grammar are common in work prepared by student writers. Learning to write well takes practice, discipline, and the careful advice of a good teacher. Despite this observation, it is possible to learn some elements of good writing from a text such as this one. In particular, understanding and avoiding common writing errors will help you immensely as you try to develop good writing habits. The writing errors listed in the following sections are typical of those found in written assignments submitted by engineering students. Review them so that you can avoid making similar mistakes in your own work. Additional examples of correct and incorrect usages can be found in references such as Strunk and White (1979).°

Parallelism.   Sentences that include multiple items or ideas should follow parallel construction.
Correct: "Our module will provide data communication, consume minimal power, and satisfy the customer's needs." (All three sentence endings begin with a verb.)
Incorrect: "Our module will provide data communication, minimal power will be consumed by it, and it will satisfy the customer's needs." (The three sentence endings don't have the same construction.)

Commas.   Use a comma to separate the second part of a sentence only when the second half could stand on its own as a complete sentence.
Correct (do use a comma): "We will supply five commands to the robot, and we will power the robot with batteries." (The second half of the sentence, "We will power the robot with batteries," is a complete sentence.)
Incorrect (don't use a comma): "We will supply five commands to the robot, and power it with batteries." (The second half of the sentence, "and power it with batteries," cannot stand on its own as a separate sentence. The comma after the word "robot" should be omitted.)

°W. Strunk and E.B. White, *The Elements of Style.* New York: MacMillan, 1979.

Past, Present, and Future Tense.   Maintain consistent tense (past, present, or future) as your writing progresses, or at least within a given paragraph.
Correct: "The routes will be difficult to change once they have been programmed into memory. This drawback also will apply to future versions of the robot." (future, future)
Incorrect: "The routes will be difficult to change once they have been programmed into memory. This drawback also applies to future versions of the robot." (future, present)

Use of the Word This.   For clarity the word "this" is best used as an adjective, not a noun or pronoun. It should be accompanied by an object of reference.
Correct: "This problem will be solved by designing a new system."
Incorrect: "This will be solved by designing a new system." (i.e., this . . . what?)

Use of the Words Input and Output.   "Input" and "output" are best used as nouns. Their use as verbs is often awkward and unprofessional.
Correct: "The input to the mixing circuit consisted of three microphone voltage signals. The output was fed to the amplifier in the form of a voltage summation."
Incorrect: "The microphone signals were inputted to the mixer. Their combined sum was outputted to the amplifier."

Punctuation Around Parentheses.   When words are set aside by parentheses, place periods *before* the trailing parenthesis if the parenthetical thought is a major part of the sentence.
Correct: "Our design project was completed on time. (We had been given a week to complete it.)
Incorrect: "Our design project was completed on time. (We had been given a week to complete it).

Infinitives ("To" Verbs).   Never split an infinitive. If you use the word "to" followed by a verb, do not put words in between.
Correct: "The purpose of this section is also to help you with your homework."
Incorrect: "The purpose of this section is to also help you with your homework."

KEY TERMS

E-mail	Memorandum	Presentation
Report	Instruction manual	

Problems

1. Write a report that outlines your vehicle design process for the Peak-Performance Design Competition introduced in Chapter 2.
2. The following document relates to the Peak-Performance Design Competition introduced in Chapter 2. It outlines the approach to be taken in the design of a system for paging contest participants over a three-minute time interval. It is an example of very poor writing.

Rewrite the proposal, taking into account the writing principles and suggestions outlined in this chapter.

> The three minute pager receiver will be based on a simple bandpass filter that is tuned to a distinct RF band for each receiver. Additionally, each receiver will tune into a general public announcement band which will broadcast voice messages or tones. The cost will be held very small by constructing our own receiver circuits.
>
> Power consumption is minimized by sleep mode. In sleep mode, the receiver's PA band amplifier will be disconnected from its power source via a relay or power monitor switch. Detection of a wakeup signal on the wakeup band will close the circuit between the PA band's amplifier and the speaker.
>
> In addition, preliminary cost research shows that a three-minute countdown circuit and LCD screen can be constructed for under nine dollars in quantities of 100. A speaker and blinking LED can also be provided at minimal cost.
>
> The countdown itself would also be initiated by reception of the wakeup signal. The end of the internal countdown would power down the PA band amplifier, or a second detection on the wakeup band would toggle the power off.
>
> The unit itself could be wearable and styled after a pager or smartcard. We have scheduled a meeting for Tuesday, January 21, at 11 am.

3. The following memo was written by an engineer responsible for designing a parts counting device. The writing style is very poor. Rewrite the memo using the guidelines discussed in this chapter.

> During our first conversation with the customer, we came to an initial design for the project and have scheduled a meeting with the clients on Tuesday. For the design, first thing come across is a detector to physically count the parts falling through the sorting mechanism and we generally prefer to use a photosensor. For the counting mechanism, two methods are proposed and yet remain undecided. One of them is to program a PLA whose reprogramming process could be too complicated for the end user. However, it's advantage is that the design would be simple and cheap. Another approach is to use a microprocessor to do the counting. However, since our team don't have any experience on this subject before, we are still seeking advice and reference. Finally, when the designated no. of parts are counted, the counter will activate a visual and audio signal which prompts the user that parts are ready for packaging. Then the user can put a plastic bag underneath the container and push a button which opens up the bottom of the container.

The above would conclude our initial idea and we will come up with more details and specifications after the meeting.

4. Write a memo to your fictitious boss asking permission to attend a technical conference.
5. Write a short instruction manual that explains how to operate your VCR.
6. Write a memo to all students using your laboratory addressing the importance of safety procedures and protocol.
7. Write a memo that summarizes the following information relating to data communication protocol:

> Data Protocol:
>
> °DCE = Data Communication Equipment (female connector)
>
> Computer, processor, host: receives data, decodes, establishes communications
>
> °DTE = Data Terminal Equipment (male connector)
>
> Terminal, printer, data board - Sends data and displays output
>
> Parallel Data: 8 or 16 bits of data sent simultaneously with a DR (Data Ready) strobe from the DTE to the DCE and a CTS (Clear to Send) signal from the DCE from the DCE
>
> Serial Data: 1 start bit; 8 data bits; 2 stop bits, 14,400 baud (bits audio); no parity
>
> Synchronous data - a clock line must be established between DCE and DTE

Asynchronous data - relies on nearly precise timing and start/stop bits

RS-232 standard (receive-send asynchronous data)

Positive and negative voltages (MARK = 1 = NEG; ZERO = 0 = POS

Held in MARK state when not in use

DB-25 Connector: pin 1 - shield

> pin 2: transmit data to DCE
>
> pin 3: receive data at DTE
>
> pin 5: clear to send (CTS) from DCE to DTE
>
> pin 7: signal ground

Note: The DTE sends data on pin 2

8. Write a proposal to your student governing body asking for money to start an on-campus amateur radio club.

9. The following memo was handed in by the employee of a small company specializing in adaptive aides for physically challenged individuals. The memo is not written particularly well. Rewrite the memo using the principles and guidelines outlined in this chapter.

> To: Xebec Management
>
> From: H. Chew
>
> This project is to work with a 47-year-old individual who has no speech capability and limited physical abilities. The subject groans and grunts to indicate discomfort, displeasure, requesting, and refusing. During dinner, our customer would like us to provide the subject with a means to indicate "I want more", "I want something else", and "I want a drink," etc.
>
> To solve this problem, we had called the customer for more information, and she would give us a video tape that is talking about the older person. We also had a team meeting to discuss the project. At the end of the meeting, we considered that we would design a box which consists of an interface panel and a data control unit. The interface panel would consist of four 2.5″ buttons. Each button represents a pre-recorded phrase. The sets of outputs from the buttons will correspond to the mode selected. The data control unit consists of the power supply, speech memory, speech synthesizer, and audio amplifier.

10. The following entries were collected by a design team working on a major software project. These notes are to be used by the team to write a summary report to the project manager indicating the features that the finished product must have. The software is to be a voice synthesizer system that will enable individuals with impaired speech capability and limited motor skills to communicate by way of a simple computer mouse. Using the rough notes provided by the design team, write the finished report to the project manager.

   - Topics covered include the alarm, requests, and greetings portions of the user interface.
   - The requests frame should be configured to allow the user to express common requests.
   - The greetings frame must have the most common greetings and be designed for easy access.
   - All frames should give the user the option to reconfigure them in any order desired.
   - The alarm frame will consist of five buttons for use in an emergency only.
   - Alarm messages include help, pain, fire, police, and ambulance.
   - Requests include drink, food, bathroom, book, pen, television, radio, and music.
   - Greetings include hello, goodbye, good morning, good evening, and good night.

11. Compose the text of an e-mail message that announces a meeting of your design team for the Peak Performance Design Competition.

12. Compose the text of an e-mail message in which you request a meeting with your boss to discuss a possible raise in salary.

13. Compose the text of an e-mail message in which you request a meeting with your professor to discuss a possible change in your course grade.

14. Compose the text of an e-mail message in which you ask a semiconductor manufacturer to send you a free sample of a microprocessor chip.

15. Compose the text of an e-mail message in which you inform a client about your travel plans for an upcoming technical review meeting.

16. Compose the text of an e-mail message in which you provide arrival information for a government contractor who is coming to your laboratory to review your work.

17. Compose the text of an e-mail message in which you ask for volunteers for a committee to review company safety standards.

18. Compose the text of an e-mail message in which you solicit volunteers to participate in the annual company blood drive for a local hospital.

19. Compose the text of an e-mail message in which you ask your boss for permission to attend the annual conference of the control and automation group of the Institute of Electrical and Electronic Engineers (IEEE).

20. Compose the text of an e-mail message in which you reassure a nervous customer that your prototype for a manufacturing system will be shipped on time.

21. Prepare a set of overhead slides in which you outline your design approach for the Peak Performance design competition described Chapter 2.

22. Prepare a set of overhead slides in which you describe the results of combustion tests on a new aircraft engine.

23. Prepare a set of overhead slides in which you outline plans for a proposed new light-rail transportation system in a metropolitan area.

24. Prepare a set of overhead slides in which you describe the benefits of a proposed new graphical user interface for a record-keeping system for a real estate company.

25. Prepare a set of overhead slides in which you describe the important features of a professional quality mountain bicycle.

26. Prepare a set of overhead slides in which you report the results of tests on a high speed data link for transferring cell phone routing information from site to site.

27. Write a letter to the human resources director of Alpha Corporation in which you reply to a classified advertisement seeking software engineers.

28. Write a letter to the human resources director of Beta Corporation in which you reply to a classified advertisement seeking entry-level mechanical design engineers.

29. Write a letter to the human resources director of Gamma Corporation in which you reply to a classified advertisement seeking biomedical engineers for synthetic drug development.

30. Write a letter to the human resources director of Delta Corporation in which you reply to a classified advertisement seeking mechanical engineers to work on developing jet engines.

31. Write a letter to the human resources director of XYZ Corporation in which you reply to a classified advertisement seeking industrial engineers to design manufacturing systems.

32. Write a letter to the human resources director of Omega State Highway Department in which you reply to a classified advertisement seeking civil engineers for highway construction.

33. Write a letter to the graduate admissions director of State University in which you request information about possible financial aid for Master's degree study.

34. Write a letter to the CEO of your company in which you highlight the details of an unethical practice that you have uncovered within the company.

35. Write a letter to the sales manager of your company in which you detail the virtues of the new pencil that your engineering division has designed.

Epilogue: The Day of the Peak Performance Design Competition

The day of the Peak Performance Design Competition has arrived, and you are set to participate in the contest. You've been working on a wedge shaped design, and you hope your slow, but strong, sturdy vehicle will be able to push faster but weaker opponents from the top of the hill. You are optimistic that your design strategy will pay off, but you are wary of several worthy opponents that are testing their cars at the vehicle preparation tables. One such vehicle, shown at the starting line in Figure 8.1, is built from lightweight epoxy board. It has large wheels, and is extremely fast.* Although this vehicle is flimsy compared to your aluminum-encased wedge, its designers have installed barbed hooks on the leading edge of the chassis. These hooks are meant to dig into the carpet-covered ramp and prevent backward motion when an opposing vehicle pushes from the front. You decide to call this vehicle "Fish Hook." Fortunately, you're not the first to compete against it, so you have ample opportunity to watch Fish Hook engage other vehicles.

You watch a match between Fish Hook and another wedge-shaped vehicle that has a shape similar to yours but is less massive. You mentally nickname this second vehicle "Lite Wedge". The designers of Lite Wedge have sacrificed wheel torque and raw mass in favor of faster speed. The judge signals the start, and several seconds later, both vehicles arrive simultaneously at the top of the hill. They engage in a bumper-to-bumper duel, but Lite Wedge does not have enough power to dislodge the front end of Fish Hook. Although Fish Hook holds its ground, it cannot make any additional headway because it's been designed to turn off power once its hooks have been set into the carpet. As shown in Figure 8.2, both vehicles are virtually equidistant from the centerline at the end of the fifteen-second time interval. The judge measures the car positions carefully with a ruler and declares Fish Hook to be the winner by a scant few millimeters. You're on the schedule later to run against Fish Hook, and you begin to question the efficacy of your wedge-shaped design. Will your slow moving vehicle be able to dislodge Fish Hook's front end and allow you to overtake the top of the hill?

Before your first match can be held, two other cars must compete on the ramp. One of them, shown on the left in Figure 8.3, is based on a design that is wedge-shaped but looks nothing like yours. Nicknamed "Molly" by its creator, this vehicle has a wedge made from a small and narrow aluminum plate. Molly has large wheels that are completely exposed to her opponent's offenses. Molly's designers have added short spikes to her wheel treads in the hope of achieving better traction. Molly's opponent, shown on

---

*The figures of this chapter show actual photos taken at the Peak Performance Design Competition at Boston University. The stories and scenarios presented are entirely fictitious. For more information about getting your own Peak-Performance design competition, contact the author at mnh@bu.edu.

Figure 8.1. Fish Hook and contestants at the startling line of the Peak Performance Design Competition. (*Photo courtesy of Boston University Photo Services.*)

the right in Figure 8.3, resembles an army tank. It uses caterpillar treads and is built on a completely exposed chassis. Molly easily dislodges "The Tank" even though the latter is first to arrive at the top of the hill. Tank's designers have included no defense mechanism to prevent the vehicle from being pushed backwards down the ramp, and it cannot maintain its position.

Your first match places you in a run against a vehicle nicknamed "The Scamper". The Scamper is an extremely fast, lightweight vehicle that zooms to the top of the hill in a few seconds while leaving behind a weight and string. When the trailing string is pulled taut against the jettisoned weight, it helps to slow the vehicle and stop it at the top of the hill. The taut string also releases a spring-loaded nail at the rear of the vehicle. This nail is designed to prevent the vehicle from being pushed backwards down the ramp.

As shown in Figure 8.4, the lightweight Scamper is no match for your massive, wedge-shaped vehicle. You easily dislodge your opponent at the top of the hill, even though you arrive there several seconds later. Your vehicle is strong enough and your wedge angle steep enough that you lift up Scamper and raise the tip of its nail right off the carpet. Your stopping mechanism places you precisely at the top of the hill and you're declared the winner of the run. Your strategy of combining a low gear ratio, slower travel speed, and higher wheel torque with the mechanical advantage of a wedge shape has paid off.

The next match places Molly against Fish Hook. The tip of Molly's narrow wedge catches under several of Fish Hook's barbs and dislodges them from the carpet. Molly's wedge is too narrow to reach them all, however, and some of the hooks retain their grasp, allowing Fish Hook to hold its position at the top of the hill. As you watch this race in action, an idea comes to mind. You decide to try adding a thin metal strip across the entire leading edge of your wedge-shaped chassis. This strip, which protrudes about 2 cm from your wedge's front end like a blunt knife, just grazes the carpet as your vehicle travels along. Hopefully, it will be able to dislodge all of Fish Hook's embedded

Figure 8.2. The judge carefully measures the positions of Fish Hook and Lite Wedge at the top of the hill. (*Photo courtesy of Boston University Photo Services.*)

hooks. You've designed your modular vehicle to accommodate rapid changes, as permitted between runs by competition rules, so the change is easy. You add the metal strip just before your run against Fish Hook. Your wide metal strip dislodges all of Fish Hook's hooks, and you dominate the top of the hill. You've made it past a formidable foe.

You watch as the wedge-shaped vehicle designed by Tina, Juan, Fred, and Karin, dubbed "Tijuana", competes against several other vehicles. Despite its clever harpoon design, Tijuana's stopping mechanism does not work as its designers had planned. In using a single switch placed beneath the chassis (see Figure 3.4 on page 61), Tina and Juan have failed to account for the height of the carpet pile. The carpet that covered their own test ramp had a much shorter pile than does the carpet covering the Peak Performance Design Competition ramp. Their stopping switch keeps snagging the ramp where its slope changes from upward to horizontal, causing their vehicle to stop short of the top of the hill. Their harpoon design works well, however, and it prevents several opponents from coming anywhere near the top of the hill. Tijuana wins several matches. Between runs, Tina and Juan solve their switch snagging problem by moving their rear wheels forward, thereby decreasing the axle-to-axle wheel span of their vehicle and leaving more room for the switch as it rounds over the transition point in the ramp. As you have done, they've designed a vehicle that is easily modified.

Tijuana next competes against Scamper. Scamper is so fast that it races to the top of the hill before Tijuana's harpoon can be launched. Tijuana is a strong wedge, however, and it easily dislodges the lightweight Scamper from the top of the hill and wins the run.

Figure 8.3.   Molly's narrow wedge approaches Tank, who has arrived first at the top of the hill. (*Photo courtesy of Boston University Photo Services.*)

Figure 8.4.   Your strong, wedge-shaped vehicle easily dislodges the lightweight Scamper. (*Photo courtesy of Boston University Photo Services.*)

You finally are paired against Tina and Juan and prepare for a wedge-against-wedge contest. As the judge bellows out, "One, two, three, go!," you engage your starting mechanism and Tina presses her starting switch. Soon after the start, Tijuana's harpoon fires, firmly lodging itself in the carpet in front of your path. You are dismayed to see the harpoon work so well, because you assume that your vehicle will not be able to dislodge its well-placed barbs. When your vehicle reaches the embedded harpoon, your newly installed knife edge dislodges the harpoon! The unseated harpoon rises over your vehicle and you continue toward the top of the ramp. You meet Tina and Juan's vehicle face to face at the top of the hill, and you each prepare for a battle of the titans. You watch as your well balanced, smooth-framed vehicle burrows under theirs, lifting it off the ramp! When Tina and Juan moved their rear wheels forward, so that their stopping switch could clear the higher carpet pile, they shifted their center of gravity toward the rear of the axle supports. This change has provided your wedge with more mechanical advantage than it might otherwise have had. You lift their front wheels off the ramp, push their wedge backwards a few centimeters, and capture the top of the hill. Your flexible design and well-conceived mid-race design modifications have lead to victory.

Index